Usage Pattern and Methodology of Indoor Gardening

Usage Pattern and Methodology of Indoor Gardening

Subhash Meena

RANDOM

RANDOM PUBLICATIONS

NEW DELHI (INDIA)

Usage Pattern and Methodology of Indoor Gardening

ISBN 978-93-5111-838-1

Published in 2016 in India by

RANDOM PUBLICATIONS

4376-A/4B, Gali Murari Lal, Ansari Road
New Delhi-110 002
Phone : +9111-43580356, 011-23289044, 011-43142548
e-mail: sales@randompublications.com,
info@randompublications.com, randomexports@gmail.com

Reprinted 2026

Type Setting by : Friends Media, Delhi-110089

Preface

An indoor garden can take up as much or as little space as you are willing to give it. Growing plants of all kinds, even tomato gardening can be done on a windowsill or on a table.

Indoor gardens benefit from a good planting medium - soil found outside is not appropriate, since it's often too heavy and may contain weed seeds and insect pests. Instead look for a mix that is specific to indoor plants. A good growing media should remain loose and drain well, yet contain enough organic matter to hold nutrients and moisture.

Indoor gardening had its first heyday during Victorian times. As plants such as abutilons, palms, and hibiscus were discovered by botanists in remote parts of the world, those who could afford to do so, filled their parlors and conservatories with these exotic beauties.

One of the major benefits of indoor gardening is that many houseplants help clean the air in the home. House paint, carpets, and building materials often release toxic gases or volatile organic compounds. Many types of houseplants are able to absorb the gases, reducing the pollution in the air. Flower gardening indoors is a way to bring color and life to a home, even in the cold winter months. Certain flowers, such as Peace Lily and Gerber daisy, also help purify the air. Growing herbs and vegetables indoors allows people to save money at the grocery store. An indoor garden of vegetables and herbs means that people can enjoy fresh, locally grown produce year-round.

This is however, an attempt to bring out a comprehensive work on indoor pants, which includes much more than shade-loving plants of ornamental foliage. On this ground the work claims to be unique.

– Author

Contents

1

Introduction

When the winter blahs set in and your dreaming of fresh greens from your summer garden, consider growing indoors. Not only do plants cleanse your household air and improve the aesthetics of any indoor space, they can provide your family with a wealth of yummy, organic foods.

City dwellers, or those without a good gardening spot in the yard, may find growing indoors especially useful. Plants don't need to take up much space — a windowsill is fine if that's all you have. For others, the indoor garden may become starter plants for an outdoor garden come spring.

GETTING STARTED

Space

An indoor garden can take up as much or as little space as you are willing to give it. Growing plants of all kinds, even tomato gardening can be done on a windowsill or on a table. Larger growers, or the more dedicated may want to set up a table or bench specifically for the garden. Find an area with a tile or linoleum floor to catch the inevitable drops of water, or place a tarp under your table. Shelves provide lots of planting room while taking up little space. If using shelves, make sure that adequate light reaches every plant. This may require a separate grow light for each shelf.

Light

Plants need light to photosynthesize and need to photosynthesize to survive. Without adequate light a plant will grow tall and spindly. If there is enough energy to grow leaves, they still may not totally expand. And without enough light, don't plan on seeing flowers or fruit. Even plants grown near a window will probably not get enough light during the winter months to thrive. There are a few things to think about when purchasing a grow light.

1. Plants have photoreceptors that absorb specific wave lengths of light. Your light needs to have the same wave lengths as the sun, which is why a regular light bulb doesn't work.

2. The light should be as close to the plant as possible without burning the leaves.
3. Most vegetables and other plants do best with 14-16 hours of sunlight or simulated light. There are a few ways you can tell if your plant is getting enough light or not. If it isn't getting enough light, it usually will have small leaves, thin stems, and the color of the plant will be lighter than usual.
4. A hormone called "florigen" controls budding and flowering. Long day plants require about 14 to 18 hours of light to produce just the right amount of florigen to flower and reproduce. Short day plants require about 10-13 hours of light. If short day plants are exposed to too many hours of light, florigen can be destroyed, preventing blooming.

SELECTING A GROW LIGHT

There are a lot of different grow lights for sale out there and it can be confusing to figure out which type is best for your indoor garden. The following run-down should bring some clarity.

Incandescent Lamps are inexpensive and can be bought at a hardware store or nursery. While they work OK for growing houseplants, they are not ideal for an indoor garden.

Fluorescent Lights work best for growing herbs and other plants that don't require a lot of light. They are not good for plants that are budding or flowering because they don't put off enough light. Inexpensive, they can be purchased at the local hardware or garden supply store.

The new Compact Fluorescent Systems, however, are quite bright and efficient and in some cases might even be better than the fancier high intensity discharge (HID) lights. Compact fluorescents are smaller and more efficient than older forms of fluorescent lighting so they can be used for all plants. They also produce less heat than incandescent and HID lights and consequently can be placed much closer to the plant.

High Intensity Discharge (HID) Bulbs are the brightest and most efficient lights available, but they can be expensive. One 1,000 watt grow light bulb can produce the same amount of light as 50 40-watt fluorescent lights.

There are several types of HID bulbs:

- High Pressure Sodium
- Metal Halide
- Low Pressure Sodium
- Mercury Vapor

The High Pressure Sodium and Metal Halide bulbs are the only ones indoor gardeners will need. High Pressure Sodium (HPS) Bulbs produce a red-orange light that benefits flowering. With an average lifespan 2X that of metal halides, high pressure sodium lamps are economical. This isn't a great light if you are only going to use one, as it doesn't produce light in the blue spectrum needed for leafy growth.

Metal Halide (MH) Bulbs produce a blue-white color that is conducive to encouraging leafy growth and keeps plants compact. A bulb will last about 10,000 hours and produce up to 125 lumens per watt compared to 39 lumens per watt for standard fluorescent lights and 18 lumens per watt for standard incandescent bulbs. This is a good light to start plants out with. When it comes time to flower, switch to a High Pressure Sodium bulb.

There is more to a grow light than just the bulb. You can purchase the reflector, cord, ballast, bulb and other parts separately, or buy a whole system that just needs to be plugged in.

What size grow light do you need? This will vary depending on the mounting height of the reflector (how far above your plants the light is) and the size of your indoor garden. In general, the following recommendations apply:

Size of Light (Wattage)	**Area Coverage Recommendations**		**Mounting Height Above Plants**
400 Watt	no outside light some sunlight	5′ x 5′ area 8′ x 8′ area	1 to 4 Feet
600 Watt	no outside light some sunlight	7′ x 7′ area 10′ x 10′ area	1.5 to 5 Feet
1000 Watt	no outside light some sunlight	8′ x 8′ area 12′ x 12′ area	2 to 6 Feet

Temperature

Temperatures of 65-75°F are best for most plants. A variance of 10°F either way will probably be OK. Plants that are too hot will be small and weak. Plants grown at too-cold temperatures may have yellow leaves that fall off.

Humidity

A lack of humidity in the house can be a challenge for indoor gardeners. Winter tends to be drier than summer, and if you run the heat in your house the problem is further compounded.

You know you have a low-humidity problem if:

1. The tips of your leaves are turning brown

2. Plants look withered or puckered
3. Plants lose their leaves
4. You've researched how much humidity your particular plant needs and it isn't getting it.

To increase humidity:

- Mist plants daily, or more often as needed. (Do not do this with hairy-leaved plants since the water hangs around longer and could cause disease.)
- Place a tray of water near your garden (don't put plants in the tray, this can lead to other problems). Fill the tray with lava rocks to increase surface area for evaporation.
- Place plants close together to create a microenvironment with a higher relative humidity.
- Run a humidifier (this might benefit your skin as well!).
- Purchase an environmental controller, which can humidify or dehumidify depending on your needs.

GROWING MEDIUM

Indoor gardens benefit from a good planting medium — soil found outside is not appropriate, since it's often too heavy and may contain weed seeds and insect pests. Instead look for a mix that is specific to indoor plants. A goodgrowing media should remain loose and drain well, yet contain enough organic matter to hold nutrients and moisture.

Most commercial organic mixes will work well, or you can create your own.

Tip: Ocean Forest Potting Soil (shown here) is a quality blend of premium earthworm castings, bat guano, and Pacific Northwest sea-going fish and crab meal. Composted forest humus, sandy loam, and sphagnum peat moss are added to provide a light, aerated texture. It's perfect for containers and ready to use right out of the bag.

HYDROPONICS

Instead of growing indoor plants in a soil mixture, you may want to try out hydroponics. Basically, this means gardening without soil. Soil holds nutrients and anchors plants roots. When growing hydroponically you provide the nutrients directly. Instead of being bound up in soil, the nutrients are readily available to the plants.

Some of the advantages of growing hydroponically include:

- Faster plant growth (up to 50% faster) since plants can easily access water and food.
- Roots grow throughout the media without becoming root bound, so containers can be smaller.

- Plants start in a disease-free medium and are less likely to become infected.
- If plants do become sick, the disease is usually in one plant, not all of them.
- Plants droop before they wilt, so you'll know to water them before they are damaged.

CHOOSING PLANTS

Almost anything can be grown indoors — as long as it eventually doesn't get too big. However, do consider growing plants with similar light, humidity and watering needs together. Some obvious choices for an indoor garden include:

VEGETABLES	
Peppers Salad Greens Kale Chard	Carrots Onions Tomatoes, especially cherry types Beans, Bush
HERBS	
Basil Parsley Oregano Lavender	Cilantro Rosemary Chives Catmint
FLOWERS	
Geranium Pansy Zinnia Roses Candytuft	Alyssum Marigold Petunia Begonia Shasta Daisy
FRUITS	
Strawberries Blueberries	Apples, dwarf varieties Citrus

Don't stop there, as mentioned above, almost anything can be grown in a container.

Plants can be grown from seed (started inside and staying inside) or they can be transplanted from your outdoor garden at the end of the season. Plants will need to be acclimated before bringing them in the house and again when you put them outside in the spring or fall.

MOVING PLANTS OUTSIDE

Plants and seedling grown inside need a period of "hardening off" before they can permanently live outdoors. The hardening off process gives them time to develop a thicker cuticle and avoid water loss while being better able to withstand the harshness of weather.

The following steps will help acclimate indoor plants to life in the great outdoors.

1. 7-10 days before you want to transplant your plants, place them outside in a shady spot or cold frame for 3-4 hours.
2. Each day, increase the time spent outdoors by 1-2 hours. Bring plants back in each night.
3. After 2-3 days, place plants in morning sun, then move them into the shade in the afternoon.
4. If the temperature stays around 50°F, plants should be able to stay out all day and night after 7 days.
5. In about 7-10 days transplant your seedlings or plants. If possible, transplant on a cloudy day and water thoroughly.

To acclimate plants by withholding water or by using a cold frame, read How to Harden Off Plants.

Moving Plants Inside

At the end of the growing season you may want to move plants inside to your indoor garden. After potting these plants (if they are not already in containers) they will need a period of acclimation, just as plants going the other direction do.

MAINTENANCE

Watering

Plants grown in containers dry out more quickly than their soil-grown counterparts and require frequent watering. Always use room-temperature water and add enough water that it runs through the drain holes of your pot or container (do not let water collect in a saucer or under the plant — this can lead to rot or disease). Use your finger to feel the soil or use a moisture meter to be sure you are not over or under watering plants.

Signs of Over Watering	Signs of Under Watering
Wilting from stem towards leaves	Wilts along the outer tips of the leaves first
Lower leaves dropping	Dry soil
Discoloration	Brown edges along the leaves
Plant might stop growing	Wilting foliage
Wilting foliage	Leaves or flowers drop prematurely

INDOOR GARDENING

HOW TO GROW BEAUTIFUL PLANTS INSIDE YOUR HOME

Most avid gardeners don't check their love of plants at the front door. We keep ferns in the bathroom, begonias in the study, a dwarf citrus in the bedroom, pots of rosemary in the kitchen, and seedlings in the basement. Today there are so many interesting plants that can be grown indoors that there's simply no reason for a gardener not to be surrounded by plants all year-round.

Fig. Multi-level plant stands are perfect for showcasing a collection of houseplants.

Indoor gardening had its first heyday during Victorian times. As plants such as abutilons, palms, and hibiscus were discovered by botanists in remote parts of the world, those who could afford to do so, filled their parlors and conservatories with these exotic beauties.

There was another indoor plant revival during the 1970s, though it was largely confined to foliage plants such as spider plants, Swedish ivy, and ferns. Today, commercial greenhouses offer an eye-popping selection of houseplants to choose from, including orchids, bougainvillea, scented geraniums, exotic begonias, pentas, and gardenias. These tropical beauties will bring both color and fragrance to your home. The trick is learning how to keep them happy indoors.

The more you know about your indoor plants and where they come from, the easier it will be to keep them healthy. Understanding a plant's native habitat can help you decide which potting soil to use, how often to water, what window the plant should be in, and whether or not it needs extra humidity. But even if you don't have specific cultural information about each and every one of your houseplants, you can feel your way to success with some general guidelines.

A FEW WORDS OF COMFORT

Plants have a lifespan, just as people do. If you have a struggling houseplant that has been around for a few years, it may simply be tired, and all the TLC in the world may not be able to revive it. Consider starting over with a new plant. Remember that unhealthy plants attract insects like a magnet, and when the infestation spreads to your other plants, you may regret your earlier large-heartedness.

Gift plants, such as cinerarias, poinsettias, chrysanthemums, azaleas and cyclamen, make wonderful indoor decorations, but in most cases they should be discarded after they have finished flowering. Some of these plants can be nursed along to flower again, but it is usually difficult to provide the growing conditions they need for another lush display of blooms.

The Right Soil

If possible, your potting soil should be tailored to the particular type of plant you are growing. Cactus, succulents and rosemary, for example, prefer a coarse, well-drained soil that is about one-third sand. Seedlings should be grown in a light, moisture-retentive, soilless mix. Pilea Plants

and ferns prefer soil with a high humus content, which can be achieved by adding leaf mold or shredded bark. Many kinds of orchids are happiest growing in nothing but fir bark or sphagnum moss.

A good indoor potting mix is usually composed of peat moss, vermiculite and perlite. These soilless mixes absorb moisture very well and resist compaction, but they tend to dry out very quickly. Since they do not contain any nutrients, you must provide your plants with a consistent supply of fertilizer. One advantage to a soilless mix is that it is sterile, so there is no chance of introducing pest or disease problems.

Many gardeners add organic components to their indoor growing mix. These might include leaf mold, finished compost, composted peat, or rich garden soil. A growing medium that contains 10 to 20 percent organic matter will usually not dry out as fast as a soilless mix, and it also allows you to introduce beneficial microorganisms and nutrients.

The most critical consideration when you're purchasing or blending your own potting soil is to ensure that the mix is light enough to provide adequate pore space for air, water and healthy root growth. Month after month of overhead watering, without the benefit of earthworms and weather to aerate the soil, usually results in an unhealthy, compacted root zone. To ensure that your plants' roots have the oxygen they need for healthy growth, your potting soil should contain plenty of perlite, vermiculite, or sharp sand. This will allow water to drain freely, and ensure that the soil is at least 10 to 20 percent air.

Climate, Water and Nutrients

Temperature: In their natural habitat, most plants experience a day-to-night temperature fluctuation of at least 10 degrees F. In your home, they will benefit from having a similar temperature differential. Most plants also expect a resting period each year; in fact, some flowering plants actually require a period of dormancy before they will set bud and flower. To simulate this resting period, you should cut back on water and fertilizer during the late fall and early winter, when the intensity and duration of natural light is lowest. Once the day length begins to increase, you can declare it to be spring, and step up the water and fertilizer. Your plants will respond with healthy new growth.

Humidity: Most plants are happiest when the relative humidity is 50 percent or higher, though they can usually survive at 30 to 40 percent. If the air is much drier than that, they are unable to absorb enough water through their roots to keep up with the water lost through their leaves. Unfortunately, indoor air, especially in the winter, often has a humidity as low as 10 to 20 percent.

Misting your plants helps, but only for an hour or so. A better solution is to use a cool vapor humidifier (which you will benefit from as well). You can also cluster your plants together so that, as they release moisture into the air, they'll humidify their neighbors. Or try arranging your plants on a gravel-filled tray that contains about 1/4 inch of water. As the water evaporates, it will humidify the air around your plants. Just be careful that the pots don't sit directly in the water.

Water: More houseplants die from overwatering than from anything else. The best advice is to resist the temptation to water on a regular schedule. Your plants will require frequent watering if the weather has been sunny and warm, and they may not need a drop if the weather has been cool and cloudy. Make a habit of checking the soil of at least a few indicator plants, and water only if it feels dry to a depth of ½ to 1 inch.

When you do water, drench the root ball until you can see some water seep out the bottom of the pot. This will ensure that the entire root ball gets moistened. Small pots will benefit from being soaked in water for about an hour, once a month. Whenever possible, try to water your plants with room temperature water to avoid shocking the roots.

Never use water that has been chemically softened. It contains salts that are harmful to plants. If your water is very hard, consider installing a demineralizing attachment to filter out impurities, such as lime and chlorine.

Fig. Plant Health Care is especially formulated for houseplants and seedlings.

Nutrients: Indoor plants are usually not too fussy about fertilizers. The most important thing is to not overdo it. Follow the instructions on the package, and err on the weak side. Always water your plants thoroughly before applying any sort of fertilizer. A standard 10-10-10 formulation is fine for most indoor plants.

Supplementing with an organic amendment such as liquid seaweed or fish emulsion, or a biostimulant, will provide some of the trace nutrients lacking in an inorganic plant fertilizer. A top dressing of compost or worm castings is another effective way to add organic nutrients. Be aware that some plants are particularly sensitive to pH level, and that this sensitivity can be either exacerbated, or corrected, with the right fertilizer. To avoid the buildup of fertilizer salts, it's a good idea to periodically drench the soil with clean water, then water again with clean water. This will help flush any salts out of the soil.

PROVIDING ADEQUATE LIGHT

Plants differ greatly in their need for light. Some will be happy with the diffused light from a north window. Others will languish if they don't receive 12 hours of bright light all year-round. Knowing the light requirements for each of your plants will allow you to determine where they will be happiest. Here are some general guidelines for matching plants to various locations in your home.

Most flowering plants, and some sun-loving foliage plants, need to be within 3 feet of a sunny, south-facing window. Plants that prefer bright, indirect light can be located 3 to 5 feet away from a south-facing window, or within 3 feet of an east- or west-facing window.

Plants that thrive in diffused light can be placed 6 to 8 feet away from a south-facing window, or within a foot of a north-facing window. In that location they'll receive about 25 percent of the light they would get if they were in front of the sunny, south-facing window.

During the winter months, you may need to move all of your plants closer to the window in order to compensate for the decrease in light. Most plants perform best when they receive 12 to 16 hours of light per day. If you want to keep your plants blooming during the short days of winter, you may need to provide supplemental lighting.

GROWING INDOOR PLANTS WITH SUCCESS (B 1318)

Much of the scenic beauty of nature has been replaced by densely populated areas that sprawl for miles from urban centers. This visual pollution affects us all and leaves us with a longing for a closer connection with nature. We spend about 90 percent of our time indoors. Interior plants are an ideal way to create attractive and restful settings while enhancing our sense of well being. In addition, houseplants can be a satisfying hobby and can help purify the air in our homes. Indoor plants not only convert carbon dioxide to oxygen, but they also trap and absorb many pollutants. Many of these chemical compounds, which are released into our air through a process called "off-gassing," come from everyday items present in our homes and offices.

To be a successful indoor gardener, you need to understand how the interior environment affects plant growth and how cultivation differs from growing plants outdoors.

FACTORS AFFECTING PLANT GROWTH

Plant growth is affected by light, temperature, humidity, water, nutrition, and soil.

Light

Of all of the factors affecting plant growth in interiors, adequate light is by far the most important. Light is needed for plants to produce food and survive — generally, the more light available, the more food produced for growth. Light is measured in units called foot candles. One footcandle (ft-c) is the amount of light cast by a candle on a white surface 1 foot away in a completely dark room. Outdoors, the light levels on a bright day range from 10,000 ft-c in an open sunny area to 250 ft-c or less in the shade of a large tree.

It is very helpful to have a general idea of how much light is present in a given location in your house. You can get a fairly good estimate with a handheld light meter, or you can use a 35 mm camera and do the following:

- Set the film speed indicator to ASA 25 and the shutter speed to 1/60th second.
- Place a piece of white paper where you want to measure the light levels, aim the camera toward the paper close enough to fill the view, and adjust the f/stop so that the meter indicates a correct exposure.
- Read the approximate light level from Table below.

Table. Indoor light levels and appropriate f/stop settings.

f/stop Setting	Light Level
f/2	40 ft-c
f/2	75 ft-c
f/4	150 ft-c
f/5.6	300 ft-c
f/8	600 ft-c
f/11	1,200 ft-c
f/16	2,400 ft-c

With the help of this table, you can obtain the light intensity reading from anywhere in your home. For example, if the f/stop setting is f/16, the approximate light level is 2,400 ft-c.

Using the light readings, your home can be divided into four areas, which have the following light levels for 8 hours per day:

1. Low-light areas: 25 ft-c – 75 ft-c
2. Medium-light areas: 75 ft-c – 200 ft-c
3. High-light areas: over 200 ft-c but not direct sunlight
4. Sunny light areas: at least 4 hours of direct sunlight

Fig. Many foliage plants are native to tropical rain forests, where light levels are low. These plants can be easily injured if exposed to strong light. Symptoms of over-exposure are upright leaves and bleached, scorched leaves. Do not place high-light sensitive plants in direct sunlight (on a porch or in front of a window). In this example, Chinese Evergreen (Aglaonema) and Dumb Cane (Dieffenbachia) show symptoms of high-light damage.

In your home, the amount of light in a given location is variable — it is affected by the presence of trees outdoors (may shade at certain times), roof overhangs (may shade at certain times), wall color (reflectance), window curtains, day length, time of day, and time of year.

When shopping for indoor plants, select plants for a given location based on the approximate light levels in the spot. The plant's label will usually contain information on the light requirements of the plant. If the plant label lists "high light" but the selected area in the home does not provide adequate light, artificial light sources such as fluorescent and/or special incandescent lights may be used to supplement the natural light.

Increasing the number of hours of light exposure can also help—for example, 16 hours of light and 8 hours of dark. This extends the number of hours during which plants receive light.

While adequate light is crucial for plant growth, too much light can be damaging.

Indoor plants are classified according to the amount of light needed for growth. Look for this information in general terms on the plant's label:

- Low: minimum 25 ft-c – 75 ft-c, 75ft-c – 200 ft-c for good growth
- Medium: minimum 75 ft-c – 150 ft-c, 200 ft-c – 500 ft-c preferred
- High: minimum 150 ft-c – 1,000 ft-c, 500 ft-c – 1,000 ft-c preferred
- Very high: minimum 1,000 ft-c, 1,000+ ft-c preferred

Windows with eastern exposure within the home generally provide the best light and temperature conditions for most indoor plant growth because plants receive direct morning light from sunrise until nearly midday. Footcandle readings at these windows can reach 5,000-8,000. As the morning progresses, the direct sun recedes from the room.

An eastern room is cooler than southern or western rooms because the house absorbs less radiant heat. Light from the east is cooler than that from the south or the west, and thus it causes less water loss from the plants.

Windows with southern exposure give the largest variation of light and temperature conditions. The low winter sun shines across the room for most of the daylight hours.

In the summer, when the sun is farther north than it is in the winter, the sun rises at a sharp angle in the morning and is high in the sky by noon. Direct light comes into a south window only at midday. If there is a wide overhang covering the windows outside, the sun may not enter the room at all. The sun at noon on a summer day may measure 10,000 ft-c. Indoors, however, a southern window with wide eaves on the outside will receive about the same amount of light as a window with northern exposure. Southern and western exposures are interchangeable for most plants. In the winter, most plants, except those with definite preference for northern exposure, can be placed in a room with southern exposure.

Windows with northern exposure provide the least light and the lowest temperature. Because the United States is in the northern hemisphere, it receives most of its sunlight from the south. Out of the four exposures, the northern exposure receives the least light and heat year round.

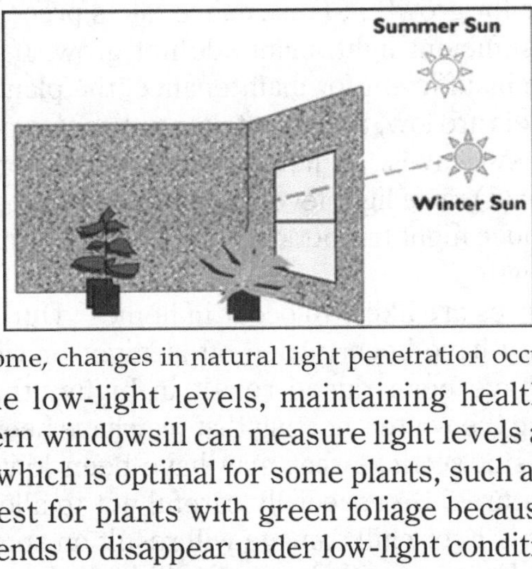

Fig. Within your home, changes in natural light penetration occur with the seasons.

Because of the low-light levels, maintaining healthy plants can be a challenge. A northern windowsill can measure light levels as low as 200 ft-c on a clear winter day, which is optimal for some plants, such as the African violet. This exposure is best for plants with green foliage because the coloration on variegated foliage tends to disappear under low-light conditions. Although most plants grown indoors will not grow in a northern room, they may tolerate it for short periods of time.

Seasons change the amount of natural light entering through windows. For example, the summer sun reaches a higher zenith compared to the winter sun. Therefore, sunlight penetrates farther into a room during winter.

How can you tell if your plant is not receiving adequate light?

- The plant does not grow.
- The internodes (spaces between the leaves) on the new growth are much longer than the internodes on the older part of the plant.
- The new leaves are smaller than the older leaves.
- The leaf color is a lighter green on the newer foliage than on the older foliage.
- The older leaves are dead.

Temperature

Temperature is the second most important factor influencing plant growth in interior environments. People feel comfortable in the range of 72 degrees F-82 degrees F, and interior plants can tolerate and grow well in the 58 degrees F-86 degrees F range because most indoor plants originate from tropical and subtropical areas of the world.

Temperature and light are linked through the processes of photosynthesis and respiration. These processes can be thought of as the "yin and yang" of plant life — two parts of a circle. Photosynthesis builds sugars and starch, which are then broken down by respiration to provide energy for the development of new tissues (growth) and the maintenance of existing ones. High temperature speeds up respiration. If the plant is not producing sufficient sugars (as under low light), then high temperatures may break down what little sugars are made, leaving little to none for growth. Maintenance takes precedence over growth; therefore, under insufficient light, plants do not grow. If light is so low that sugars produced are insufficient for maintenance, the plant eventually dies.

When sugar levels are low, the plant takes nutrients and sugars from older leaves to maintain new leaves. To help plants in an indoor environment, two options are available: (1) raise light levels to increase photosynthesis and sugar production or (2) reduce night temperature to lower respiration rates and allow more sugars for growth.

What temperatures are likely to occur in homes? During the summer, air conditioning that may have been turned off at night or weekend thermostat settings that may have been raised result in higher than desirable night temperatures. During the winter, heating that may have been turned off at night or weekend thermostat settings that may have been lowered may result in lower night temperatures. Be especially careful not to allow temperatures to drop below 50 degrees F, or chill damage will result on some sensitive foliage plants (e.g., Chinese Evergreen, Aglaonema). Chill damage is manifested with the yellowing of lower leaves and/or defoliation.

Plants vary in their minimum and maximum temperature requirements. Examples of cool-loving plants suitable for locations where temperatures drop to the low 50s at night and 60s during the day are Cyclamen, Wonder Plant, Fatshedera, Japanese Aralia, and Fatsia. A list of plants and their temperature requirements. Not all interior plants have the same temperature requirements for optimal growth. For example, Cast Iron Plant, Aspidistra, and ferns actually grow better with cooler temperatures (72°F), while other tropical plants grow best if the temperatures are 90 degrees F – 95 degrees F. Such temperatures are rarely allowed indoors. The best temperature range for indoor plants is 70 degrees F – 80 degrees F day and 65 degrees F – 70 degrees F night.

Relative Humidity

Relative humidity is the amount of moisture contained in the air. For interior plants, relative humidity below 20 percent is considered low, 40 percent – 50 percent is medium, and above 50 percent is high. Relative humidity is a very important factor, but it is easily overlooked. In a greenhouse, relative humidity is 50 percent or higher. Rapid transpiration and water loss may result when newly purchased plants are placed in the 10 percent – 20 percent relative humidity typical of most homes. Most indoor plants come from the tropics where high relative humidity is common. Therefore, take the following steps to help your plants adjust to the low relative humidity in your home.

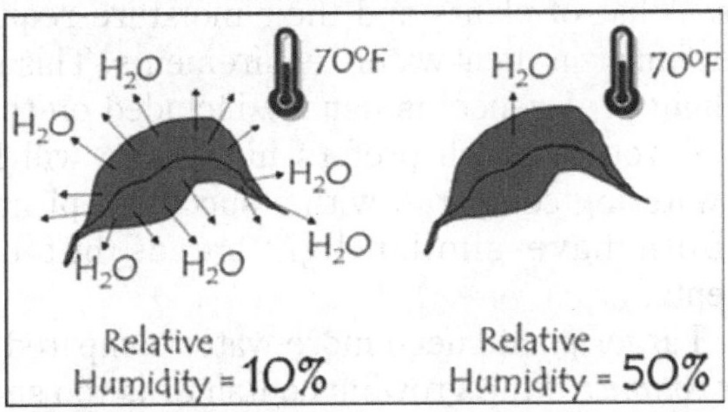

Fig. The lower the humidity, the more water is lost from a leaf. At the same temperature of 70°F, a leaf placed in 10% relative humidity loses more water compared with the leaf placed in 50% relative humidity.

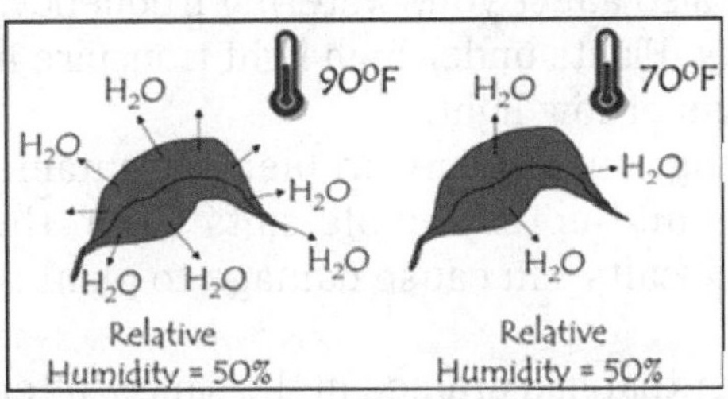

Fig. The higher the temperature, the more water vapor the air can hold, and the more water will be lost from the plant. At the same humidity of 50%, a leaf placed in 90°F air loses more water compared with the leaf placed in 70°F air.

- Place plants close together to create a microenvironment with a higher relative humidity.
- Use a shallow container filled with water and lava rocks or gravel, which will provide evaporation from a large surface area and increase relative humidity.
- Use a humidifier.
- Use mist bottles to spray water around the plant; however, in reality, you would need to mist every few minutes for an indefinite amount of time to make a difference in relative humidity around the plant.
- The foliage and flowers of plants with hairy leaves should not be sprayed with water. Water on such leaves may stay longer, providing opportunities for disease spores to germinate.

WATER

Water Quantity

Learning to water is one of the most important skills in plant care. Applying too much water can suffocate plant roots and too little water causes growth to become erratic and stunted. Watering frequency will depend on the conditions under which the plants are growing. When dealing with how much water to apply, consider the following:

- *Plant type:* A list of plants and their moisture requirements. Not all plants are similar in their water requirements. This information, along with the light preference, is usually included on the plant label. For example, a croton, which prefers high light, will likely need more frequent watering compared with a succulent plant such as Opuntia cactus. Both have similar light needs but dissimilar water requirements.
- *Plant size:* Larger plants need more water compared to smaller plants.
- *Container volume:* If the growing container is too small, watering may be required more frequently.
- *Soil moisture:* The amount of water already present in the growing medium will also affect your watering frequency.
- *Light intensity:* Plants under high light transpire more water compared with plants under low light.

Improper watering causes many problems. Containers with saucers may cause an excessive build-up of soluble salts (from the applied fertilizer). High levels of soluble salts can cause damage to plant roots and a decline in growth.

Discard any water that had drained in the saucer after irrigation, and apply large quantities of water to the soil to leach the accumulated soluble salts. In deciding when you should water, feel the soil by pushing a finger an inch or so below the surface. If the soil is still moist, no further water is needed. Water devices or water meters are also available to simplify watering.

Water Quality

The quality of the irrigation water is an issue with plants that are susceptible to fluorine and chlorine, such as Corn Plant (Dracaena), Ti Plant (Cordyline), Peacock Plant (Maranta), and Rattlesnake Plant (Calathea). Alleviate this problem by letting the water stand for several days — so that some chlorine and fluorine will be released from it — before applying the water to the plants. Move susceptible plants away from the edge of the pool to prevent water splashes from reaching the foliage. Do not use susceptible plants around enclosed pools. In general, plants with long linear leaves (such as the Spider Plant) are more susceptible to fluorine.

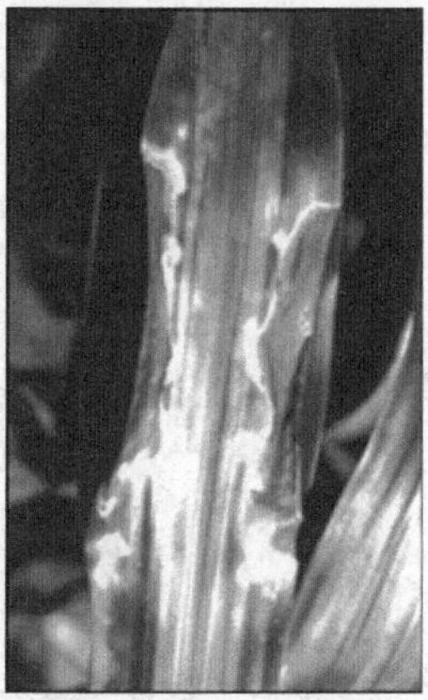

Fig. Symptoms of fluorine damage on the Corn Plant include tip and leaf scorching.

Fig. Symptoms of fluorine damage on the Ti Plant include tip and leaf scorching.

NUTRITION

Many indoor gardeners have the same problem with fertilizer that they have with water — they want to give their plants too much. Danger from over-fertilization occurs because any fertilizer used, whether in liquid, powder, or tablet form, will dissolve in soil water and will form salts in the water. When you over-fertilize, the water in the soil becomes so salty that it "burns" the plant's roots by removing water from them. Excess soluble salts accumulate as a whitish crust on the surface of the growing medium and/or near the rim of the container.

Fig. Soluble salt burn is manifested as leaf martinal and tip burn. Dead roots also invite root diseases.

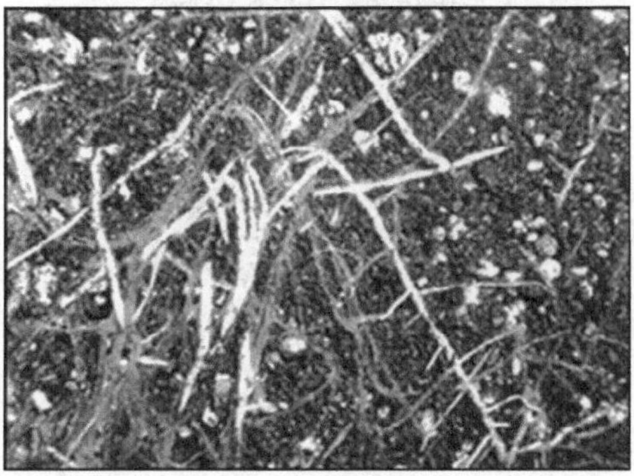

Fig. Soluble salts can burn roots; notice that the healthy roots are white, while the dead roots are brown. Dead roots also invite root diseases.

Before feeding plants, consider the following:

Plant type: Some plants are heavy feeders (e.g., Ficus species), while others need little or no additional fertilizer for months (e.g., succulents).

Volume of soil: The growing medium that is present — smaller pots require less fertilizer compared with larger pots because they contain less soil.

Light intensity: The higher the light levels, the more nutrients needed for plant growth.

A newly purchased, healthy plant rarely needs an immediate application of fertilizer. In most cases, the amount of fertilizer applied by the commercial producer will supply enough nutrients for two to three months in the home. This rule is flexible — if deficiency symptoms are evident, fertilizer application is desirable.

The secret to fertilizing plants indoors is to apply small amounts of fertilizer as the plant grows. Without new growth, the plant has a limited need for more fertilizer. During the winter when light levels are low, a plant's need for fertilizer reduces. During the summer when light levels increase and the plant is actively growing, its need for fertilizer increases. As a starting point, use about one-fourth the label rate for monthly applications. If the overall plant color becomes lighter green, fertilize every two weeks. If the new growth is dark green but the leaves are small and internodes seem longer than on the older growth, decrease the fertilizer rate.

Varying fertilizer formulations are available to the indoor gardener. Many fertilizers come in specially designed formulas for indoor plants. Generally, they contain a lower percentage of the required mineral elements to prevent over-fertilization problems.

Soil/Growing Medium

The growing medium provides anchorage, water, and minerals. When repotting plants, make sure that the new mix is well drained and aerated, holds water and nutrients well, and is within the right pH range (5.0-6.5). A good potting mix provides ample amounts of oxygen to the root system. Most professional mixes are good to use. Some plants require special mixes, e.g., bromeliads, orchids, and African violets. Either purchase these mixes or prepare your own. Below are some formulas that can be used to prepare a homemade potting mix.

GROWING MIX FOR FLOWERING HOUSE PLANTS

The following potting mix will grow acceptable flowering plants in most homes for most gardeners:

- 1 part garden loam or potting soil
- 1 part sand or perlite or vermiculite
- 1 part peat moss

Add 2 to 3 ounces of 20 percent superphosphate and ¾ ounce of either bonemeal or dolomitic limestone (by weight) to 4 gallons of potting mix. After sterilizing the soil, add 3 tablespoons of a 6-6-6 or similarly balanced fertilizer to every 4 gallons (½ bushel) of mix. Add a minor element formulation according to the manufacturer's recommendations.

Growing Mixes for Foliage Plants

Although most foliage plants will grow satisfactorily in the growing mix recommended for flowering house plants, they will grow better if the mix contains a higher percentage of organic matter.

- 1 part garden loam or potting soil
- 1 part sand or 2 parts peat moss

OR

- 1 part pine bark
- 2 parts peatmoss

OR

- 1 part sand
- 1 part pine bark
- 1 part peat moss

Add 2 to 3 ounces (dry weight) of dolomitic limestone to 4 gallons (½ bushel) of mix. For fluoride-sensitive plants, adjust the pH so it is no lower than pH 6.5. Superphosphate contains enough fluoride to cause foliar burn on sensitive plants. After sterilizing the soil, add 3 tablespoons of a 6-6-6 or another fertilizer such as 5-10-5 to each ½ bushel. Plastic-coated fertilizers also can be used; most of them require about 2 ounces per ½ bushel. Add a minor element formulation to the potting mix per the manufacturer's recommendation.

Growing Mixes for Bromeliads

Bromeliads are plants from Central and South America, which are either epiphytic (they grow on tree branches or in the crotches of trees) or terrestrial (they grow in the ground). Although most of the bromeliads can be grown successfully in foliage plant mixes, most grow better in specially designed soil mixes. Any mix for bromeliads must be well aerated and drained.

- 2 parts peat moss
- 1 part perlite
- 1 part fir bark

OR

- 1 part peat
- 1 part pine bark

OR

- 1 part peat
- 1 part pine bark
- 1 part cypress shavings

Add 2 ounces of dolomitic limestone to 4 gallons (½ bushel) of soil mix and a minor element mix. Dissolve 1 ounce of 10-10-10 water-soluble fertilizer in 3 gallons of water. Use this solution after repotting and again monthly when watering. Also, add enough water to fill the vase formed by the overlapping leaf bases.

Growing Mixes for Orchids

Orchids have a great deal in common with bromeliads because they also grow on trees as epiphytes and on the ground as terrestrials. A mix for orchids should have excellent drainage and aeration, too. Some soil mixes that can be used are:

- 3 parts osmunda tree fern fiber (moisten before use by soaking in water for 12 hours)
- 1 part redwood bark

OR

- 5 parts fir bark
- 1 part perlite

Tree fern slabs may also be used to grow epiphytic orchids.Add 1 ounce (dry weight) of dolomitic limestone per 4 gallons (½ bushel) of soil mix. Do not add fertilizer to the mix. After the plants are potted, add ¼ ounce of liquid 10-10-10 with minor elements per gallon of water and fertilize once every 6 weeks (if the plants are growing in osmunda fern fibers). If plants are growing in fir bark, use a liquid 30-10-10 with minor elements every 6 weeks instead of a 10-10-10 fertilizer.

Growing Mix for Succulents and Cacti

Cacti and other succulents grow best in a well-drained and aerated soil.

- 2 parts garden loam or potting soil
- 2 parts sand
- 2 parts peat
- 1 part perlite (crushed charcoal can be substituted)

Add 2 ounces (dry weight) of dolomitic limestone to 4 gallons (½ bushel) of soil mix, 2 ounces (by weight) of bonemeal, and ½ ounce of superphosphate. After sterilizing the soil, add a minor element supplement according to the manufacturer's recommendation.

Growing Mix for Ferns

Ferns grow well in most recommended mixes that have a high proportion of organic matter with good soil aeration and drainage characteristics. Use any of the suggested foliage plant mixes. However, most ferns kept indoors grow better in the following mix:

- 1 part garden loam or potting soil
- 1 part peat moss
- 1 part pine bark
- 1 part coarse sand

Add 2 ounces (dry weight) of dolomitic limestone to each ½ bushel (4 gallons) of soil mix and ½ ounce of either bonemeal or 20 percent superphosphate. After pasteurizing the soil mix, add minor elements to the mix. Add 1 tablespoon of a 6-6-6 or similarly balanced fertilizer to each ½ bushel of soil mix.

GROWING MIX FOR AFRICAN VIOLETS

Any number of soil mixes for African violets exist, and most of them will grow high quality plants. A good mix should be well drained and aerated.

- 2 parts peat moss
- 1 part vermiculite
- 1 part perlite

Add 2½ tablespoons of dolomite and 1½ tablespoons of 20 percent superphosphate to each ½ bushel of soil mix. Add 3 tablespoons of a high phosphorous fertilizer such as 5-10-5 or a similar ratio of fertilizer.

How to Sterilize Soil

Sterilization reduces the number of diseased organisms and weeds present in the soil. First, mix the soil with an equal portion of vermiculite or peat moss (otherwise, the soil will become very hard). Next, moisten the mixture and place it in the oven. Allow it to "bake" at 180 degrees F–200 degrees F for 1 hour. Once the soil cools, it is ready to use. To treat soil in the microwave, first mix the portion with an equal amount of vermiculite or peat moss and moisten. Place the mixture in a plastic bag. Next, consult the manufacturer's manual to determine the amount of time and power level needed to heat the quantity of soil to about 180 degrees F (most portions of soil will generally require about 10 to 15 minutes). Insert a probe into the soil and make sure it has heated to 180 degrees F – 200 degrees F. Allow the soil to cool before using it or storing it for future use.

Make sure that the soil or potting mix you want to sterilize does not contain perlite. At high temperatures, toxic levels of fluoride may be released and subsequently damage your plants.

Acclimatization

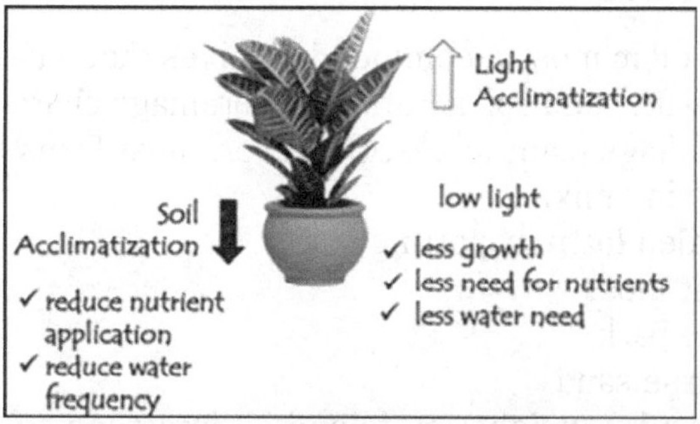

Fig. The two sides of acclimatization — the above-ground (light acclimatization) involves adaptation to low light, which means less growth and less need for nutrients. The below-ground (soil acclimatization) involves reduction of nutrients and water.

Acclimatization is the adaptation of a plant to a new environment, and it is very important for the health and growth of indoor plants. In greenhouses, plants are accustomed to high light, nutrition, water supply, temperatures, and relative humidity — conditions ideal for fast growth. Residential homes, with low-light interiors and low relative humidity, will most likely produce a stressful experience for plants — the greater the difference between the previous environment and the environment of the house, the greater the stress the plant endures.

Acclimatization is generally done in the greenhouse or the nursery. Plants are grown for a period of time under low-light levels and with fewer nutrients. Because this slows down plant growth, acclimatized plants are not ready for the market as early as nonacclimatized plants. Acclimatized plants cost more compared to nonacclimatized plants, but this is money well spent. The symptoms and appearance of acclimatized plants.

Fig. Acclimatized Weeping Fig; notice the large, dark green leaves and the elongated internodes.

To acclimatize plants at home, place newly purchased plants in bright areas for at least 3 to 4 weeks and then move them to their final location. Porches and patios are ideal bright places for your plants in the warm months, as long as the plants are not in direct sunlight. The most common symptom occurring in plants placed indoors is defoliation. As long as it is not extensive and it slows down after a few weeks, the plants will adjust to the particular location. Keep in mind, however, that each time the plant is moved around, it will experience an acclimatization period, and such changes may become evident. Learn as much as possible about the extent of acclimatization of the chosen plants. The retailer should be able to provide this information. When shopping for plants at a garden center, ask if the plants have been acclimatized. Remember that the most important factors of indoor plant growth are adequate light, fertilizer, and water at reduced rates.

Table. Symptoms of acclimatized plants vs. nonacclimatized plants.

Acclimatized Plants	**Nonacclimatized Plants**
Medium to dark green leaves	Yellowish to light green leaves
Large leaves	Small leaves
Flat leaves	Partially folder leaves
Thin leaves	Thick leaves
Widely spaced leaves	Closely spaced leaves
Long internodes	Short internodes
Thin to medium stems	Thick stems
Horizontal or slightly flexed leaf position	Upright leaf position
Few new leaves	Many new leaves
Wide branch angles	Acute angles

WHAT TO LOOK FOR WHEN SHOPPING FOR INDOOR PLANTS

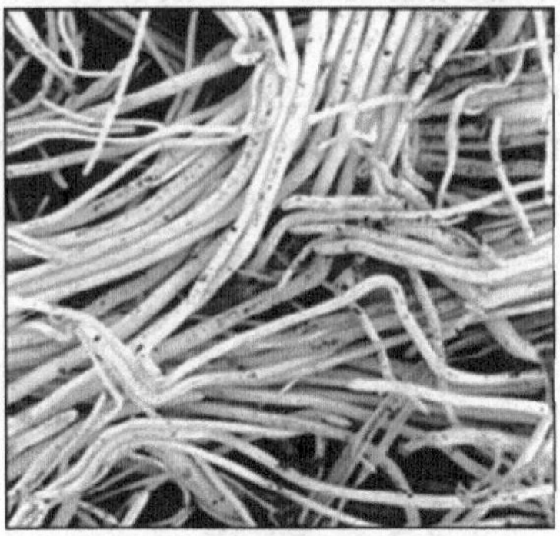

Fig. Healthy roots are typically white without any discolorations.

Purchase only healthy looking plants with medium to dark green foliage (unless foliage is supposed to be a different color).

Avoid plants with unnaturally spotted, yellow, or brown leaves. If the plant is unhealthy at the nursery, chances are that it will die soon after consumer purchase. Look for pests on the undersides of leaves. Remove the plant from the pot and examine the root system. Healthy roots generally are and should be visible along the outside of the soil ball and should have an earthy smell.

Any discolorations, generally brown or blackened roots, are signs of problems. Some plants, such as Dracaenas, have roots with colors other than white. Unhealthy roots also may smell foul. If shopping for ferns, do not be alarmed if you see brown-colored spots or long rows of structures on the lower leaf surface; these "spots" are reproductive structures called spores.

SELECTING CONTAINERS

Planters can enhance the decorative value of the plants. Consider the following when selecting a planter:

- Suitability for the plant's needs
- Suitability for the needs of the individual and the environment
- Cost and availability
- Strength and durability
- Drainage
- Weight

The style, shape, and size of the container should complement the plants grown. Small containers are best for small slow-growing plants, while fast-growing plants are better suited for large containers.

Containers can be made from a wide range of materials — terra cotta, clay, plastic, or ceramic. Terra cotta pots, made of fired clay, are some of the most popular choices, with designs ranging from plain to ornate. Plants perform very well in terra cotta pots, as the porous surface allows good air exchange between the plant roots and the environment. Other clay containers (not considered terra cotta) range from gray to brown in color, depending on the clay used. Clay pots can be glazed or unglazed. The glazed pots restrict air exchange but offer more design choices. Unglazed pots evaporate water faster and plants in them may need more frequent watering. Disadvantages of clay containers include their weight (especially large pots) and the chance they will chip or break.

Constructed of materials such as polyethylene, polyurethane, recycled plastic, and fiberglass, plastic pots have evolved from very simple to quite elaborate. They have the advantage of being lightweight as well as chip- and break-resistant. Air exchange and water evaporation rates are generally lower in plastic containers compared with clay containers. Plants in plastic pots will not dry out as quickly as plants in clay pots, increasing the danger of overwatering.

In general, there are two types of containers — ones with drainage holes and ones without. Do not allow plants in containers with drainage holes to sit in saucers filled with water, unless the plant is suspended above the water level by a layer of rocks. To avoid salt buildup, leach the soil once a month by applying a gallon of water to every cubic foot of potting medium; after a few hours, follow with ½ gallon of water. If the potting medium contains garden soil, apply 5 gallons of water per every cubic foot of growing medium.

Containers without drainage holes work well for plants such as the Peace Lily (Spathiphyllum), which needs plenty of water, but they should not be used for cacti and succulents.

PRUNING, GROOMING, CLEANING, AND REPOTTING

When is the best time to prune? "When the knife is sharp" goes the old saying, and it means using the natural life cycles as a guide. For example, when the plant is growing rapidly and you want to maintain a certain size, prune lightly and frequently, removing shoots or shoot tips when they are small. When removing the very immature tips, the practice is known as pinching. Pinching and light pruning also increase branching of the stem and result in a stockier, fuller plant.

When the plant has outgrown its container, root pruning is advisable. Pull roots away from the root mass then cut them back to within 1 inch of the soil mass. An alternative method is to make three or four vertical cuts 1 inch deep in the soil ball on the opposite sides of the root ball.

If you are re-using containers, make sure that they are clean by washing out any old compost, chemical, or paint residues. Sterilize the container by placing it in a 10 percent bleach solution and rinse well.

A clean plant is a healthy plant. Water flow causes salt accumulation along the leaf margins and/or tips, creating necrotic areas. Dust dulls normal leaf coloration, lessening plant value, but it also shades plant surfaces, reflecting light that can be used in photosynthesis. Dust on lower leaf surfaces may clog stomata (specialized cells involved in water transpiration), inhibiting gas exchange within the leaf. Leaves with thick, shiny cuticles (Croton, Ficus, Peace Lily, Bromeliads) should be cleaned with a damp sponge.

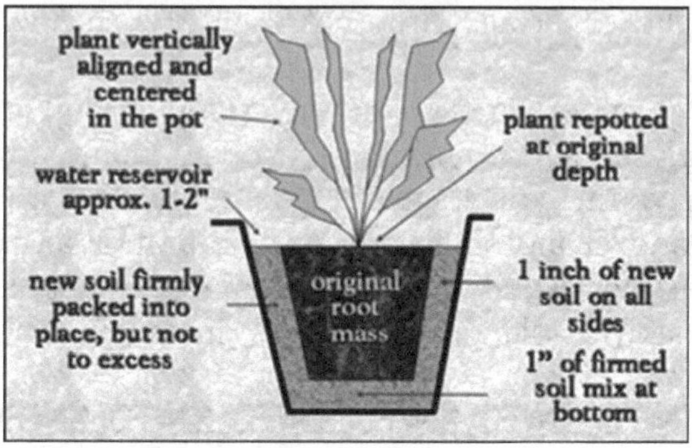

Fig. Properly repotted plant.

If the plant is small, dip the foliage in tepid water and swirl it around. Water should not be used when cleaning cacti, African violet leaves, and other plants with hairy leaves. Instead, use a clean, small paintbrush brush to remove dust. Remove dead flowers and leaves regularly. Leaves with tip and/or marginal necrosis, such as fluoride damage, should be trimmed to the healthy part.

If the plant has been growing well, it will likely need repotting. The decision to repot should be based on plant appearance—if it is top-heavy, if it fills the container with new shoots, or if it has extensive root growth out of the pot's drainage holes. Ideally, plants should be repotted in 1 inch increments. Planting into too large a container will give the roots more soil than they initially need. The excess soil will hold extra moisture, creating overly wet conditions. Increase pot size through smaller increments rather than doubling the pot size in one step.

PEST MANAGEMENT

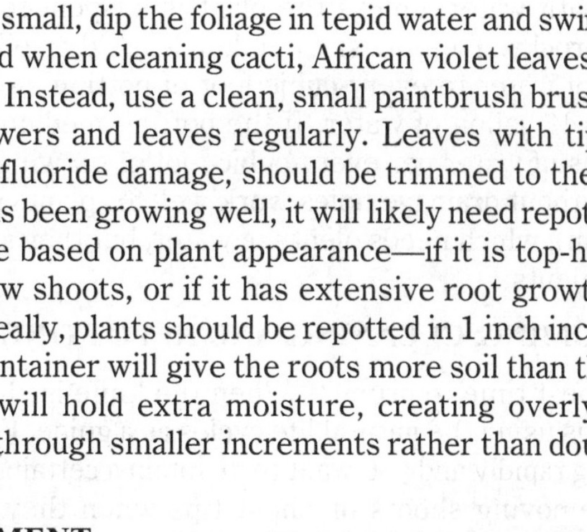

Fig. Various scales.

Very few plants stay pest-free forever. Pest insects are more likely to be encountered on indoor plants than diseases because the interior environment rarely offers favorable conditions for foliar diseases to develop. However, when plants are grown under stressful conditions (such as low light and excess water), soil-borne pathogens often develop.

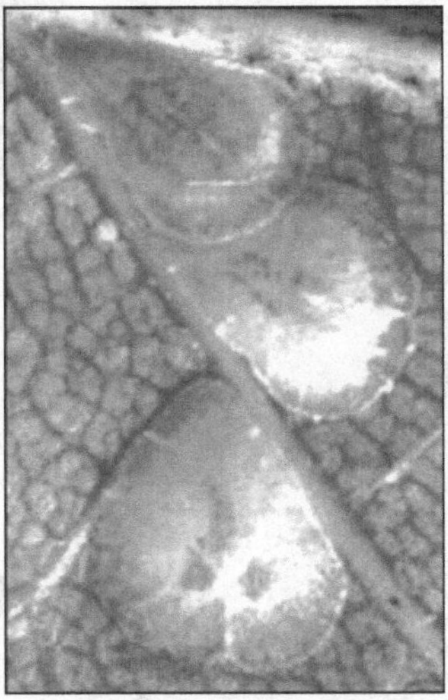

Fig. Various scales.

Scales are 1/8 inch to 1/3 inch long with various colors, depending upon the species. The three main families of scales are armored (the body covering can be separated from the body), soft (the body covering cannot be separated from the body), and mealybugs. Scales suck plant juices from leaves and stems, causing stunting, leaf discoloration, and death of the tissue. As a result of their feeding, sticky "honeydew" (digested plant sap) is excreted (the exception is armored scales). Honeydew offers a growing medium for a fungus called sooty mold, which, when present, can detract from the plant appearance and block light from reaching the leaf surface. Scales are usually inconspicuous; by the time infestation is noticed, the population is usually very large.

Mealybugs are soft bodied, 1/5 inch to 1/3 inch long, and covered by white, waxy filaments, giving them a white, cottony appearance. Insects are frequently found on the new growth at the stem apex, where they suck plant juices, causing leaf wilting and abscission. Some species of mealybugs appear first on the undersides of leaves. Mealybugs excrete sticky honeydew, which attracts sooty mold.

Aphids are soft bodied, pear shaped, 1/25 inch to 1/8 inch long, and are usually green in color (but may be pink, blue, brown, yellow, or black). Aphids reside on new growth or on the underside of young leaves, where they suck plant juices, causing deformed, curled growth of new leaves, buds, and flowers. Aphids also excrete honeydew. Aphids are usually wingless but develop winged forms when colonies become too large.

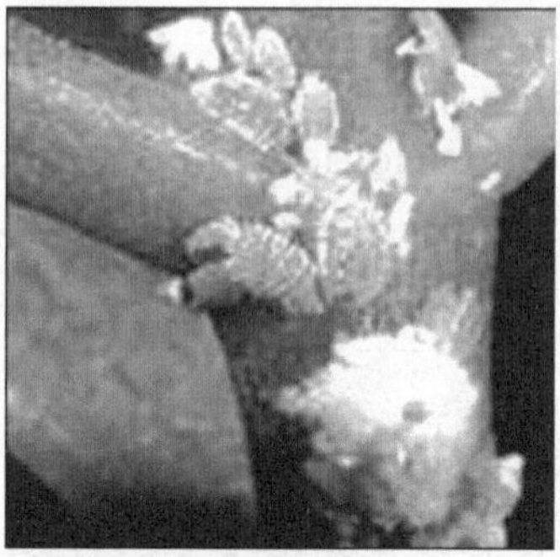

Fig. Mealybugs are the major pest problem for houseplants.

Fig. Aphids.

Spider mites are the second most common pest problem on houseplants. The adult females are about 1/50 inch long, hardly visible with the unaided eye. Mites feed on the undersides of young leaves. Infected areas are grayish or yellow speckled.

Webs form as a means of dispersal. Spider mites thrive in hot and dry conditions.

Fig. Spider mites with webbing.

Fig. Thrips.

Thrips, while uncommon on houseplants, predominantly feed on plants in patios and other outdoor areas. Thrips are small, slender, 1/25 inch to 1/12 inch long, and tan, black, or brown in color, with lighter markings. Adults and larvae feed on shoot tips, flowers, and leaves by sucking sap and cell contents. Injured tissue has a whitish or silver-flecked appearance due to the light reflecting from the empty cell walls of the dead cells.

WHAT TO DO FOR PLANT PROBLEMS

Pests

- The best method is prevention — purchase pest-free plants.
- Remove a light infestation of mealybugs or aphids with a cotton swab dipped in rubbing alcohol.
- If outdoor conditions permit, take the affected houseplant outside in a protected area, where natural predators will eventually come and rid the plant of the pest.
- Treat with insecticidal soap. The best results occur on plants that have been hardened off in the interior environment. New plants, if they have not been acclimatized (accustomed to lower light, fertilizer, and water levels), are going to be tender and should be treated after the first couple of weeks. Add 2 teaspoons of insecticidal soap per gallon of water and wipe foliage and stems with the soapy water and soft cloth.
- Heavy infestations may be too extensive to treat. Discard these plants and do not place them in your compost pile.
- Do not introduce beneficial insects indoors! They may work great in the greenhouse with a large number of plants and pests, but there is just not enough food in your home to sustain their population. Most pests can be controlled culturally on indoor plants without the use of chemicals.

Another potential problem in the indoor garden is the occurrence of various diseases. For a disease to happen, three factors must be present: (1) a susceptible plant, (2) a viable pathogen, and (3) a favorable environment. Because the home has very low relative humidity and water is often applied directly to the growing medium (thus keeping the foliage dry), chances of a foliar disease occurring are minimal.

Leaf spots are the most common problem, but they are usually not caused by a disease. For example, leaf scalds occur when water droplets on the leaves act as lenses and focus excessive light in one spot, bleaching the chlorophyll and killing the underlying tissue. Spots with patterns are signs of a disease, including a tan center, dark borders, and/or light-colored borders called "halos." Dark structures may be present on the underside; these contain a means of dispersal called spores.

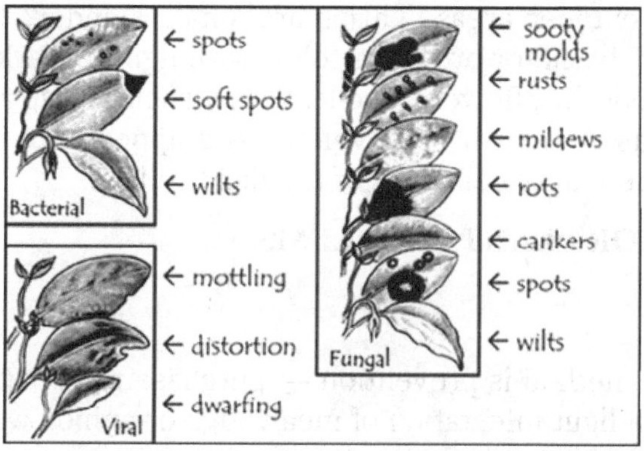

Fig. Bacterial diseases cause spots, soft spots, and wilts. The signs of a fungal disease are sooty molds, rusts, mildews, rots, cankers, spots, and wilts. Viral diseases cause mottling, distortion, and dwarfing.

Most importantly, avoid causing stress to plants. A healthy plant is much more likely to fight off a disease than a stressed one. Use a simplified key for identifying the causal agent for a disease.

Soil-borne pathogens are commonly found on stressed plants. Soil-borne pathogens affect plants at or below the soil line; disease development is usually well underway before symptoms are noted on plant parts aboveground. Soil-borne diseases commonly occur when the growing medium is kept excessively moist and fertility levels are high. Low light and over-watering create favorable environments for soil-borne diseases indoors. The most common causes of stress in interiors are low light and over-watering.

Summary of Cultural Care

A listing of more than 200 plants and their cultural requirements. To summarize cultural care guidelines, the following abbreviations and coding numbers are used. These guidelines apply to actively growing indoor plants.

L = Light
1. Sunny light areas: At least 4 hours of direct sun
2. High-light areas: Over 200 ft-c, but not direct sun
3. Medium-light areas: 75 ft-c to 200 ft-c
4. Low-light areas: 25 ft-c to 75 ft-c

T = Temperature
1. Cool: 50°F night, 65°F day temperatures
2. Average: 65°F night, 75°F day temperatures
3. Warm: 70°F night, 85°F day temperatures

H = Relative Humidity
1. High: 50% or higher
2. Average: 25% to 49%
3. Low: 5% to 24%

W = Watering
1. Keep soil mix moist
2. Surface of soil mix should dry before re-watering
3. Soil mix can become moderately dry before re-watering

S = Suggested Soil Mix

For specific ingredients, refer to the various growing mixes in "Soil/ Growing Medium." The soil mixes are keyed as follows:
1. Flowering house plants
2. Foliage plants
3. Bromeliads
4. Orchids
5. Succulents and cacti
6. Ferns
7. African violets and other Gesneriads

Table. Indoor plants and their cultural requirements (adopted from McConnell, D. B. 1978)

Botanical Name	Common Name	Cultural Care				
		L	T	H	W	S
Abutilon hybridum	Flowering Maple	1	1	2	2	1
Acalypha hispida	Chenile Plant	1	2	2	2	1
Achimenes hybrids	Magic Flower	2	2	2	1	7
Acorus calamus	Sweet Flag	2-3	2	2	1	2
Acorus gramineus	Miniature Sweet Flag	2-3	2	2	1	2
Adiantum raddianum	Maidenhair Fern	2-3	2	1	1	6
Adromischus cristatus	Crinkle-Leaf Plant	2-3	2	2	2	5
Adromischus festivus	Plover Eggs	2-3	2	2	2	5
Aechmea fasciata	Silver Vase	2-3	2	2	2	3
Aechmea miniata 'Discolor'	Purplish Coral Berry	2-3	2	2	2	3
Aechmea 'Royal Wine'	Royal Wine Bromeliad	2-3	2	2	1	3
Aeschynanthus marmoratus	Zebra Basket Vine	2	2	2	1	7
Aeschynanthus pulcher	Lipstick Vine	2	2	2	1	7
Agave Americana 'Marginata'	Variegated Century Plant	1	2	3	3	5
Agave victoriae-reginae	Queen Agave	1	2	2	2	5

Aglaonema modestum	Chinese Evergreen	3-4	2	2	2	2
Aglaonema 'Silver King'	Silver King	3-4	2	2	2	2
Aglaonema 'Silver Queen'	Silver Queen	3-4	2	2	2	2
Allamanda cathartica	Allamanda	1	2	1-2	2	1
Alloplectus nummularia	Miniature Pouch Flower	2-3	2	1-2	1	7
Aloe aborescens	Candelabra Plant	1	3	3	3	5
Aloe barbadensis	Medicine Plant	1	3	3	3	5
Aloe brevifolia	Brevifolia Aloe	1	3	3	3	5
Ananas comosus	Pineapple	1-2	2	2	1	3
Anthurium clarinervium	Dwarf Crystal Anthurium	2-3	2	1-2	1	2
Anthurium hookeri	Bird's Nest Anthurium	2-3	2	1-2	1	2
Anthurium scherzeranum	Flamingo Flower	2-3	2	1-2	1	6
Aphelandra squarrosa	Zebra Plant	2	2	2	1	2
Araucaria heterophylla	Norfolk Island Pine	2-3	2	2	1	2
Ardissa crenata	Ardisia	2-3	2	2	1	2
Asparagus densiflorus 'Myers'	Plume Asparagus	2-3	2	2	2	2
Asparagus densiflorus 'Sprengeri'	Foxtail Fern	2-3	2	2	2	2
Asparagus falcatus	Sickle Thorn	2-3	2	2	2	2
Aspidistra elatior	Cast Iron Plant	3-4	2	3	2	2
Asplenium daucifolium	Mother Fern	3	2	2	1	6
Asokebuyn budys	Bird's Nest Fern	3	2	2	1	6
Astrophytum myriostigma	Bishop's Cap	2	2	3	3	5
Beaucarnea recurvata	Ponytail	1	2	3	3	5
Begonia cubensis	Cuban Holly	2-3	2	2	2	2
Begonia metallica	Metallic Leaf Begonia	2-3	2	2	2	2
Begonia x rex-cultorum	Rex Begonia	2-3	2	2	2	2
Begonia semperflorens	Wax Begonia	1-2	1	2	2	1
Billbergia nutans	Queen's Tears	2-3	2	2	2	3
Billbergia pyramidalis	Urn Plant	2-3	2	2	2	3
Billbergia zebrina	Zebra Plant	2-3	2	2	2	3
Bougainvillea spp.	Bougainvillea	1	2	3	3	1
Brassaia actinophylla	Schefflera	2-3	2	2	2	2
Brassaia arboricola	Dwarf Schefflera	2-3	2	2	2	2
Caladium spp.	Caladium	2	2	1	1	2
Calathea insignis	Rattlesnake Plant	2-3	2	2	1	2
Calathea makoyana	Peacock Plant	2-3	2	2	1	2
Calathea micans	Miniature Maranta	2-3	2	2	1	2
Calathea roseopicta	Rose Calathea	2-3	2	2	1	2
Calceolaria crenatiflora	Slipperwort	2	1	1	1	1
Callisia elegans	Striped Inch Plant	2-3	2	2	2	2
Carissa grandiflora 'Bonsai'	Bonsai Natal Plum	1-2	2-3	2	2	1
Carissa grandiflora 'Boxwood Beauty'	Boxwood Beauty	1-2	2-3	2	2	1
Caryota mitis	Fishtail Palm	2-3	2	2	2	2
Catharanthus roseus	Madagascar Periwinkle	1-2	2	1-2	2	1
Cereus peruvianus	Peruvian Apple Cactus	1	2-3	3	3	5
Ceropegia woodii	Rosary Vine	2-3	2	2	2	5
Chamaedorea elegans	Parlor Palm	3-4	2	2	2	2
Chamaedorea erumpens	Bamboo Palm	3-4	2	2	2	2
Chamaerops humilis	European Fan Palm	2-3	2	2	2	2
Chirita lavandulacea	Hindustan Gentian	2-3	2	1-2	1	7
Chlorophytum comosum 'Variegatum'	Variegated Spider Plant	2-3	2	2	1	2
Chlorophytum comosum 'Vittatum'	Spider Plant	2-3	2	2	1	2
Chrysalidocarpus lutescens	Areca Palm	2-3	2	2	1	2
Chrysanthemum morifolium	Chrysanthemum	1	2	2	1	1
Cissus antarctica	Kangaroo Vine	2-3	2	2	2	2
Cissus rhombifolia	Grape Leaf Ivy	2-3	2	2	2	2
Cissus rotundifolia	Wax Cissus	2	2	3	3	2

Cissus striata	Miniature Grape Ivy	2-3	2	2	2	2
Citrofortunella mitis	Calamondin Orange	1-2	1	2	2	1
Clivia miniata 'Grandiflora'	Kafir Lily	2	2	2	2	1
Codiaeum variegatum	Croton	1	2	1	1	2
Coffeaa arabica	Coffee	2	2	2	2	1
Coleus blumeri	Coleus	2-3	2	2	2	1
Colummea hybrids	Goldfish Plant	2-3	2	1-2	1	7
Cordyline terminalis	Ti Plant	2	1-2	2	2	2
Crassula argentea	Jade Plant	2-3	2	2	2	2
Crassula falcata	Propeller Plant	1-2	2	2	3	5
Crassula hemisphaerica	Arab's Turban	1-2	2	2	3	5
Crassula lycopodioides	Toy Cypress	1-2	2	2	2	5
Crassula schmidtii	Red Flowering Crassula	2-3	2	2	2	5
Crassula teres	Rattlesnake Tail	2-3	2	3	3	5
Anthurium species						
	Crossandra	2	2	2	1	1
Cryptanthus bivittatus 'Minor'	Dwarf Rose Stripe Star	2	2	2	2	3
Cryptanthus fosteranus	Stiff Pheasant Leaf	2	2	2	2	3
Cryptanthus zonatus	Zebra Plant	2	2	2	2	3
Cyrtomium falcatum 'Rochfordianum'	House Holly Fern	2-3	2	2	2	6
Davallia fejeensis	Rabbit's Foot Fern	2-3	2	1	1	3
Dieffenbachia 'Exotica Perfection'	Exotica Perfection	2-3	2	2	2	2
Dieffenbachia maculata	Spotted Dumb Cane	3	2	2	2	2
Dizygotheca elegantissima	False Aralia	2-3	2	2	2	2
Dracaena deremensis 'Janet Craig'	Janet Craig	2-4	2	2	2	2
Dracaena deremensis 'Warneckii'	Warneckii	2-4	2	2	2	2
Dracaena fragrans 'Massangeana'	Corn Plant	2-3	2	2	2	2
Dracaena marginata	Marginata	2-4	2	2	2	2
Dracaena surculosa	Gold Dust Dracaena	2-4	2	2	2	2
Dyckia brevifolia	Miniature Agave	1-2	2	3	2-3	2
Dyckia fosterana	Silver and Gold Dyckia	1-2	2	3	2-3	3
Echeveria agavoides	Molded Wax	1-2	2	3	3	5
Echeveria elegans	Mexican Snowball	1-2	2	3	3	5
Echinocereus reichenbachii	Lace Cactus	1-2	2	3	3	5
Epidendrum atropurpureum	Spice Orchid	2	2	1-2	1	4
Epiphyllum hybrids	Orchid Cacti	2	2	2	2	1
Epipremnum aureum	Golden Pothos	2-4	2	2	2	2
Epipremnum aureum 'Marble Queen'	Marble Queen	2-4	2	2	2	2
Episcia cupreata	Flame Violet	2	2-3	1	1	7
Episcia dianthiflora	Lace-Flower Vine	2	2-3	1	1	7
Episcia reptans	Scarlet Violet	2	2-3	1	1	7
Euphorbia coeralescens	Blue Euphorbia	2-3	2	2-3	2-3	5
Euphorbia mammillaris	Corncob Cactus	1	2	2-3	3	5
Euphorbia milii splendens	Crown-of-Thorns	1	2	2-3	3	5
Euphorbia pulcherrima	Poinsettia	1-2	2	2	2	1
Euphorbia tirucalli	Milkbush	1-2	2	2	2	1
Fatshedera lizei	Botanical Wonder Plant	2-3	1-2	2	2	2
Fatsia japonica	Japanese Aralia	3-4	1-2	2	2	2
Ficus benjamina	Weeping Fig	1-3	2	2	2	2
Ficus deltoidea	Mistletoe Ficus	2-3	2	2	2	2
Ficus elastica 'Decora'	Rubber Plant	1-3	2-3	2	2	2
Ficus lyrata	Fiddle-Leaf Fig	1-3	2	2	2	2
Ficus pumila 'Minima'	Dwarf Creeping Fig	2-3	2	2	2	2
Ficus retusa	Cuban Laurel	2-3	2	2	2	2
Ficus sagittata	Rooting Fig	2-3	2	2	2	2
Ficus willdemaniana	Dwarf Fiddle-Leaf Fig	2-3	2	2	2	2
Fittonia verschaffeltii	Red-Nerved Fittonia	2-3	2	1	1	2

Fittonia verschaffeltii argyroneura	Silver-Nerved Fittonia	2-3	2	1	1	2
Fuchsia hybrida	Fuchsias	2	1-2	1	1	1
Gasteria hybrida	Ox Tongue	2	2	2	3	5
Graptopetalum amethystinum	Jewel Leaf Plant	2-3	2	2-3	3	5
Guzmania lingulata 'Major'	Scarlet Star	2	2	1	2	3
Guzmania monostachia	Striped Torch	2	2	1	2	3
Gynura aurantiaca 'Purple Passion'	Purple Passion	2-3	2	2	2	2
Haworthia cuspidata	Star Window Plant	1-2	2	3	2-3	5
Haworthia fasciata	Zebra Haworthia	1-2	2	3	2-3	5
Haworthia subfasciata	Little Zebra Plant	2	2	3	2-3	5
Haworthia truncata	Clipped Window Plant	1-2	2	3	3	5
Hedera canariensis	Algerian Ivy	2-3	1-2	2	2	1
Hedera helix	English Ivy	2-3	1-2	2	2	1
Hemigraphis alternata	Waffle Plant	2-3	2	2	2	2
Hibiscus rosa-sinensis	Chinese Hibiscus	1	2	2	2	1
Hippeastrum hybrids	Amaryllis	2	2	2	2	1
Howea belmoreana	Belmore Sentry Palm	3-2	2	2	2	2
Howea forsterana	Kentia Palm	2-4	2	2	2	2
Hoya carnosa 'Variegata'	Wax Plant	2-3	2	2-3	2	2
Hoya kerrii	Sweetheart Hoya	2	2	2	2	2
Hyacinthus orientalis	Hyacinth	2	1-2	2	1	1
Impatiens wallerana 'Variegata'	Busy Lizzie Impatiens	2-3	2	2	2	1
Ixora coccinea	Ixora	1	2	2	2	1
Jatropha integerrima	Peregrian	1	2	2	2	1
Justicia brandegeana	Shrimp Plant	1-2	2	2	2	1
Kalanchoe blossfeldiana	Christmas Kalanchoe	1-2	2	2	2	1
Kalanchoe pumila	Dwarf Purple Kalanchoe	1-2	2	2-3	3	5
Kalanchoe tomentosa	Panda Plant	1-2	2	2-3	3	5
Malvaviscus arboreus	Turk's Cap	1	2	2	1	1
Mammillaria bocasana	Powder Puff	1-2	2	3	3	5
Manettia inflata	Firecracker Plant	2	2	1-2	2	1
Maranta leuconeura erythroneura	Red Nerve Plant	2-3	2	2	2	2
Maranta leuconeura kerchoviana	Prayer Plant	2-3	2	2	2	2
Mikania ternata	Plush Vine	2-3	2	2	2	2
Monstera deliciosa	Philodendron Pertusum	2-4	2	2	2	2
Monstera obliqua	Window Leaf	3	2	2	2	2
Nautilocalyx lynchii	Black Alloplectus	2-3	2	2	1	7
Neoregelia carolinae 'Tricolor'	Tricolor Bromeliad	2-3	2	2	2	3
Neoregelia spectabilis	Fingernail Plant	2-3	2	2	2	3
Neoregelia zonata	Zonata	2-3	2	2	2	3
Nephrolepis exaltata 'Bostoniensis'	Boston Fern	2-3	2	1-2	2	6
Nephrolepis exaltata 'Fluffy Ruffles'	Fluffy Ruffles	2-3	2	1	2	6
Nidularium innocentii nana	Miniature Bird's Nest	2-3	2	2	1	3
Opuntia vilis	Little Tree Cactus	1-2	2	3	3	5
Opuntia vulgaris	Irish Mittens	1-2	2	3	3	5
Oxalis flava	Finger Oxalis	1-2	2	2	2	1
Oxalis rubra	Red Oxalis	1-2	2	2	2	1
Pachyphytum oviferum	Pearly Moonstones	1-2	2	2-3	2-3	5
Pachystachys lutea	Yellow Shrimp Plant	2-3	2	2	2	1
Paphiopedilum hybrids	Ladyslipper Orchids	2-3	2	2	1-2	4
Pedilanthus tithymaloides'Variegatus'	Devil's Backbone	2-3	2	2	2	5
Pelargonium hortorum	House Geranium	1-2	1-2	2-3	2	1
Pelargoniium peltatum	Ivy Geranium	1-2	1-2	2	2	1
Pellaea rotundifolia	Button Fern	2-3	2	2	1-2	6
Pellionia pulchra	Satin Pellionia	2-3	2	2	1-2	2
Pentas lanceolata	Egyptian Star Cluster	1	2	2	2	1
Peperomia caperata	Emerald Ripple	2-3	2	2	2	2

Peperomia crassifolia	Leather Peperomia	2-3	2	2	2	2
Peperomia obtusifolia	Baby Rubber Tree	2-3	2	2	2	2
Philodendron bipennifolium	Fiddle-Leaf Philodendron	3-4	2	2	2	2
Philodendron 'Emerald Queen'	Emerald Queen	2-4	2	2	2	2
Philodendron 'Florida'	Florida	2-4	2	2	2	2
Philodendron scandens oxycardium	Heart-Leaf Philodendron	2-4	2	2	2	2
Philodendron selloum	Selloum	2-4	2	2	2	2
Phoenix roebelenii	Pigmy Date Palm	2-3	2	2	2	2
Pilea cadierei	Aluminum Plant	2-3	2	1-2	1	2
Pilea microphylla	Artillery Plant	2-3	2	1	1	2
Platycerium bifurcatum	Staghorn Fern	2-3	2	2	2	6
Plectranthus australis	Swedish Ivy	2-3	2	2	2	2
Plectranthus australis	Candle Plant	2-3	2	2	2	2
Podocarpus macrophyllus	Podocarpus	2-3	2	2	2	2
Polyscias balfouriana 'Marginata'	Variegated Balfour Aralia	2-3	2	2	2	2
Polyscias fruticosa	Ming Aralia	2-3	2	2	2	2
Rhapis excelsa	Lady Palm	2-3	2	2	2	2
Rhododendron hybrids	Azaleas	2	1-2	1	1	2
Ruellia graeciznas	Red-Spray Ruellia	1-2	2	2	2	1
Saintpaulia hybrids	African Violets	2-3	2	2	1	7
Sansevieria parva	Parva Sansevieria	2-3	2	3	2-3	5
Sansevieria trifasciata 'Hahnii'	Birdsnest Sansevieria	2-4	2	3	2-3	5
Sansevieria trifasciata 'Laurentii'	Gold-Banded Sansevieria	2-4	2	3	2-3	5
Saxifraga stolonifera	Strawberry Geranium	2-3	1-2	2	2	2
Schlumbergera bridgesii	Christmas Cactus	2-3	2	2	2	2
Schlumbergera truncata	Christmas Cactus	2-3	2	2	2	2
Scindapsus pictus	Silver Pothos	3	2	2	2	2
Sedum spectabile	Showy Sedum	1-2	1-2	2-3	2-3	5
Sempervivum arachniodeum	Cow Web Houseleek	1-2	1-2	2-3	2-3	5
Setcreasea pallida 'Purple Heart'	Purple Heart	1-2	2	2	2	2
Sinningia speciosa	Gloxinia	2-3	2	1-2	2	7
Soleirolia soleirolii	Baby Tears	2-3	2	1-2	1	2
Spathiphyllum 'Clevelandii'	Peace Lily	2-3	2	2	1	2
Spathiphyllum 'Mauna Loa'	Mauna Loa	2-3	2	2	1	2
Stapelia nobilis	Carrion Flower	1-2	2	2-3	2-3	5
Streptocarpus x hybridus	Cape Primrose	2-3	2	2	2	7
Strobilanthes dyeranum	Persian Shield	2-3	2	2	2	2
Syngonium podophyllum	Nephthytis	2-4	2	2	2	2
Trillandsia bulbosa	Dancing Bulb	2	2	2	2	3
Tillandsia lindenii	Blue-Flowered Torch	2	2	2	2	3
Tolmiea menziesii	Piggyback Plant	2	1-2	2	2	2
Tradescantia blossfeldiana	Flowering Inch Plant	2-3	2	2	2	2
Tradescantia sillamontana	White Velvet	2-3	2	2	2	2
Vriesea splendens	Flaming Sword	2	2	2	2	3
Yucca elephantipes	Spineless Yucca	2	2	3	2	2
Zebrina pendula	Wandering Jew	2-3	2	2	2	2

2

Benefits of Indoor Gardening and Houseplants

One of the major benefits of indoor gardening is that many houseplants help clean the air in the home. House paint, carpets, and building materials often release toxic gases or volatile organic compounds. Many types of houseplants are able to absorb the gases, reducing the pollution in the air. Flower gardening indoors is a way to bring color and life to a home, even in the cold winter months. Certain flowers, such as Peace Lily and Gerber daisy, also help purify the air. Growing herbs and vegetables indoors allows people to save money at the grocery store. An indoor garden of vegetables and herbs means that people can enjoy fresh, locally grown produce year-round.

SELECTING CONTAINERS

Indoor gardening means that the plants need to be placed in a container of some sort. Gardeners aren't limited to the flower pots and window boxes sold at most garden centers. Any type of container will suffice, as long as it has certain features. The most important feature for any flower gardening container to have is drainage holes either on the side or bottom. Without the holes, the soil and roots of the plants will become waterlogged and can potentially rot.

The size of the container depends on the size of the plant. A pot that is too small will quickly smother a plant while a pot that is too large won't drain adequately, leading to rotting. A general rule of thumb is to pick a pot that is just an inch or so wider than the plant's original root ball. How long the plant is expected to last also determines the size of the container. An annual herb or flower won't grow very much or develop an extensive root system and will be alright in a smaller pot.

Material is another consideration when selecting a container. Certain materials, such as plastic and glazed ceramic, retain water better than others. Unglazed clay or terra- cotta pots quickly draw water from the soil, making them a better pick for succulents or plants that thrive in dryer conditions. The weight of the material is also something for an indoor gardener to consider. If they plan on moving the plant frequently, a lightweight container made of plastic might be a good option.

SELECTING PLANTS

Low maintenance plants, such as jade and dragon tree, are ideal for growing indoors. If a gardener wants to grow vegetables indoors, they should look for plants that don't need much light or space. Greens such as spinach and lettuce work well indoors, as they thrive in shady areas, don't need much space, and can be cut down and will grow back. Smaller herbs such as basil, thyme and chives are also good picks for indoor gardening. Annual flowers such as pansies, marigolds, and daisies grow well indoors, as long as they have adequate light. Most houseplants thrive in an indoor setting, as they don't need direct sunlight and are suited for the typical indoor temperatures.

Plant Growth Factors

All plants need some amount of light to survive in an indoor garden. Some plants need sufficiently more light than others. Usually, if a plant produces flowers or fruits, it needs more light than plants that only produce leaves. If a house has several large windows that face south or east, that might provide enough light for the indoor gardener. If not, the gardener should place the plants under a fluorescent lamp to give them enough light.

Vegetables, flowers and houseplants do best in temperatures between 60 degrees and 75 degrees Fahrenheit, the typical temperature range for most homes. Too-warm temperatures will cause the plants to become leggy. Different plants need different levels of humidity. Ferns and tropical plants need high humidity conditions and might thrive in a place that is naturally more humid, such as the bathroom. Succulents such as jade prefer drier conditions.

Indoor plants need to be fed and watered, too. A fertilizer high in nitrogen is ideal for plants grown for their leaves while fertilizer with a higher amount of phosphorous is ideal for flowering plants. Plants that don't get enough fertilizer stop growing, develop pale or yellow leaves, and stop producing flowers or fruits.

One issue many indoor gardeners have is giving their plants too much water. Houseplants and indoor vegetables should only be watered when the soil is completely dry, at least one inch deep. An easy way to water the plants is to pour water over the soil until it flows out of the drainage hole on the underside of the container. The plants can be watered from below, too, by placing a saucer full of water underneath the pot.

Plant Care and Maintenance

Plants grown indoors should be fertilized when they are in a period of active growth, usually from March to October. A gardener can feed the plants about once a month, following the package directions. The plants don't need feeding when they are dormant during the colder months of the year. Even though the plants are indoors, they still become dormant for several months. In the winter, indoor plants also need less water.

Once plants outgrow their containers, they should be moved to a bigger pot. A plant is too big for its container once its roots start growing out of the drainage hole or once water passes through the pot without saturating the soil. A gardener should pick a pot that is an inch or so larger than the initial pot. When repotting, it's important that a gardener only use container mix or potting mix. Soil from the garden isn't appropriate for a container, as it will become compacted and retain water. Garden soil can also contain pests and pathogens.

INDOOR PLANTS: NATURE'S AIR FRESHNERS

Most homes have at least one house plant. Possibly the most popular is the low-maintenance cactus. House plants serve to soften the hard interior and create interest by inviting nature into the home. Other benefits from house plants include higher humidity, added color and healthier air.

Aromatic plants bring into a room or house an often overlooked benefit. Not all house plants are fragrant, but some have a pleasant scent and thrive indoors. The long winter months, when only the smell of cooking fills the indoor air, is a good time to introduce some new fragrances that can help bring a glimpse of spring into the home, says Jerry Goodspeed, Utah State University Extension horticulturist.

"Fragrant plants can help mask cooking and other odors, eliminating the need for air fresheners," Goodspeed says. "It's preferable to have a nice plant emitting a fragrant aroma than have the smell of burnt toast masked with the smell of 'tree in a can.'" He says scented geraniums are probably the most popular fragrant house plant. They do not thrive indoors, but will stay healthy if placed in a well lit location, plus there is a wide assortment of aromas. Among the most popular are lemon, almond, pine trees, peppermint, orange, pineapple and even chocolate.

However, one of the drawbacks with some scented geraniums is their scent is weak, Goodspeed explains. Some require that you rub the leaves to get a strong whiff of their fragrance. This would be fine, but most plants can only stand so much rubbing before they die. "It's the foliage of scented geraniums that carries the pleasant aroma," he says. "These geraniums need to dry out between watering, and be fertilized only about once a month during the winter. In the summer they can be moved outdoors and allowed to bloom."

Goodspeed says another blooming aromatic house plant is the gardenia. Lack of light is the biggest problem with getting a gardenia to grow and bloom indoors. They require a sun room, and do best in indirect light. In the winter they need to stay cool at night (55 to 60 degrees), and can be a high maintenance plant. He suggests other flowering plants that may be easier to grow in the house: pink jasmine (Jasminum polyanthum), string of beads (Senecio rowleyanus), wax flower (Stephanotis floribunda), and wax plant (Hoya). These plant all require some special care and bloom better in indirect light in a sun room or other well-lit location. They all should be allowed to slow down and rest during the winter, by reducing the fertilization and keeping them somewhat cooler.

Goodspeed says the Hoya, string of beads and wax flower are all vining plants and can be trained up a wire or trellis, or planted in a hanging basket. The flowers have a sweet fragrance that can be smelled throughout a small greenhouse or large room. "The common hyacinth (Hyacinthus orientalis) can be planted in a pot and forced to bloom indoors during the winter months," he says. "It also has a very sweet aroma that can fill a room. After they have bloomed and the foliage has died back, they can be planted outdoors and enjoyed for many springs to come. Buy new bulbs each year for forcing indoors."

HOUSEPLANTS: CONTAINERS

Choosing the proper plant container is very important. Select containers that allow drainage, or place a layer of gravel or broken pot shards in the bottom of the pot to allow drainage. Ideally, plant containers should have a drainage hole in the bottom. If not, soil can become waterlogged and root rot may occur.

Decorative containers often do not have drainage holes, so it may be necessary to double-pot your plant. Double potting allows you to enjoy a decorative container while providing proper drainage for your plant. When double potting, put a small amount of gravel in the bottom of the decorative container and place the clay orplastic container inside.

Porous ceramic pots allow more air to reach plant roots than plastic, but have a tendency to show stains from salts in water.

When using tall, narrow pots, you'll need to use a finer textured soil to maintain even moisture than when using short, wide pots. Larger pots require less frequent watering. Decorative baskets may also be used, but you'll need to use a plastic saucer inside the basket to protect furniture and floors from damage resulting from drainage.

Select containers according to plant type. For example, succulents may do better in unglazed clay containers while moisture-loving plants, like Boston fern, may do better in plastic containers. Glazed clay, wire with peat moss, and wooden containers are also available. Avoid treated wood containers because they may affect plant growth and viability.

CHOOSING A CONTAINER FOR PLANTING

Anything that holds soil and has drainage holes in the bottom may be transformed into a container garden for terrestrial plants. If it doesn't have drainage holes or they can be plugged, then see Constructing and Caring for Water Gardens. Consider eye appeal, convenience, and cost, when choosing a container. Also think about how easy it will be keep the plants healthy. For vibrant plant growth, the containers must provide adequate space for roots and soil media, allowing the plant to thrive. Use containers of vigorous plants to

- Provide focal points
- Divide outdoor rooms
- Create privacy
- Screen objectionable views
- Accent the landscape
- Grow tasty herbs, fruits and vegetables.

Your choice of container (along with choice of plants) communicates feelings and sets a mood. It can be amusing, a statement of value, clever, loud, quiet, classy, creative, solid, sophisticated, stylish, primitive, homespun, environmentally friendly, understated, matched and more. Larger plants may require heavier and stronger pot materials to prevent the pot from blowing over or keep the roots from breaking the pot.

USING SOIL AND SOIL MIXES

Well prepared garden soil is great for growing things in the ground but when it comes to growing things in containers, soil as you know it needs to be changed. Soils for containers need to be well aerated and well drained while still being able to retain enough moisture for plant growth. When choosing what to use to fill containers, never use garden soil by itself no matter how good it looks or how well things grow in it out in the garden. When put into a container both drainage and aeration are severely impeded, and the results are that plants grow poorly or not at all.

Soils for containers are always modified in some way to ensure proper drainage and aeration. Container soils are often referred to as soilless or artificial media, because they contain no soil at all. They are often composed of various things such as peat, vermiculite, bark, coir fiber (ground coconut hulls) in a variety of recipes depending on the manufacture and the type of plant material being grown. They can be found under a variety of trade names and in sizes ranging from a few quarts to bales that are many cubic feet in size. Sometimes the choice of media will be directed by what type of plants you are growing. Succulents, herbs, and perennials tend to prefer soils that are well drained and not retaining a lot of moisture over a long period of time. For them you might choose media that are courser in texture containing more bark, perlite or sand. For tropicals and foliage plants, you might choose a media with more peat and less course material as these plants tend to prefer moisture growing conditions.

When these mixes are used, they should be moistened slightly before planting. Fill a tub with the media, add water and lightly fluff the media to dampen it. Garden soil can be used as a container media but it needs to be modified or amended. An acceptable soil based mix can be made by using one part garden soil, one part peat moss and one part perlite or coarse builders sand. Don't use fine beach sand or play sand.

This media has advantages and disadvantages. The disadvantage is that the soil you are using may contain insects, weed seeds and disease organisms. Soilless medias are generally free of these things. Soil based medias are heavier. This may be an advantage though especially if the containers are used in a very windy location and the extra weight will help keep the pot upright. Soil-based medias are also a bit more forgiving when it comes to water and fertility. They tend not to dry out as fast, and they also tend to hold on to nutrients longer. Because soilless medias can be a bit more costly, you could mix 25 percent soil with the soilless media to stretch your soilless media for a few more pots.

Also, can soilless media be reused from year to year? To a certain extent, yes. If the plants in the containers were healthy during the growing season with no major disease issues, you could remove the plant material at the end of the season and reuse the media next season. You may have to add some additional new media to fill the container. There will come a time when the media in the container will need to be replaced. Over the course of time, the organic materials that the soilless media is made from break down and decompose to the point where you will loose the drainage and aeration properties that are inherent in soilless container medias. When that happens, discard the media to the compost pile or to the garden and refill the container with fresh media.

There are also bagged products labeled as top soil that tend to be largely sedge peat. While they are inexpensive and look very good, once put into a pot they are poorly drained and poorly aerated. They can be used in combination with soilless media products or amended much the same way as if you were using garden soil.

When filling containers with media, don't fill the pot to the top. Leave about a one inch space between the top of the soil and rim of the pot. This will help make watering the pot easier as it provides a place to "put water" and not have it run over the edge.

Filling very large containers can be costly especially when using commercially prepared media. To reduce the cost and also the weight of the container consider adding "filler" to the bottom of the container to take up space. Many things can be used, but they should be something that is inert, able to take up space and not break down over the course of the growing season. While there are inserts made for this purpose, there are a lot of home products that can be recycled for this use.

Items such as crushed aluminum cans, plastic milk jugs, and non-biodegradable "packing peanuts" are usually readily available. Fill the bottom one-quarter to one-third of the container with your choice of material. Lay a piece of landscape fabric over the top of the material and fill the rest of the container with media.

The filler takes up space the landscape fabric keeps the soil from infiltrating the filler while allowing water to pass through, and there is still ample space for roots to grow.

CONSIDERING ENVIRONMENTAL FACTORS

Light and Temperature

Light and temperature are key environmental factors to successfully growing and maintaining plants whether they are in the ground or in containers. Containers offer the advantage of being portable. As the seasons, temperature and light conditions change, you can move your containers to maintain the desired conditions for peak performance.

It is important to know the preferred conditions for the plants you grow, whether they are patio tomatoes that need long sunny days or big leaf hostas that can sunburn in bright light. If relocating containers is not a possibility, or a desired option, then changing the plants to match the temperature and light conditions can be a successful strategy.

Wind

Wind may be a seasonal problem causing containers to blow over. Be sure the combined weight of the container and soil is proportional to the upper portions of the plant.

Plants with large leaves can be damaged in locations with strong wind. Flowers and buds can become desiccated and hence, short-lived. One way to address these issues is to relocate the container.

Heat Absorption

When considering the light conditions in your planting area, don't overlook the absorption of heat by dark-colored containers. Dark-colored containers exposed to full summer sun can get so hot that roots are damaged and the potting media dries out very quickly. Move dark-colored containers to shaded areas, or shade the container with trailing plants to reduce heat absorption. Another strategy for dark-colored containers is to fill them with Mediterranean type plants such as herbs and succulents that can handle hot root zones and prefer dry soils.

CREATING FOUR SEASON CONTAINERS

Containers offer four seasons of pleasure, fun and creativity. Make use of this opportunity to be creative and plan on changing out the plants in your containers frequently during the growing season. Check out Blueprints for Building Containersfor additional ideas and photos. Gorgeous containers bursting with colorful arrangements are a focal point and should always look their best.

The amount of time and money you want to spend on your containers will determine how many times each year you want to change the plantings or add additional plants. It may mean completely changing some pots with each season or just changing a few plants in each pot. Or it may mean simply rearranging your containers so that the best looking ones are always front and center, while those containers that are not at their peak are in the background.

Spring

A feeling of seasonality can be expressed by changing color schemes or textures. Spring planting schemes can include bulbs, pansies, violas, dianthus, primulas, lobelias, snapdragons and fillers such as ivies and other foliage plants.

Summer

In summer, large containers can be planted with tropical plants using bulbs such as caladiums, bananas, and cannas for a tropical feel. Large containers can also be planted with miniature sunflowers, ornamental millets, zinnias and celosia to create a Midwestern feel in a sunny location. Summer flowering plants for smaller containers include Dahlberg daises, zinnias and dahlias.

Fall

Fall color schemes revolve around oranges, deep golds and rich reds. Mums are the classic standby, but calendulas, pansies, ornamental kales, diascias, snapdragons and edibles such as beets and Swiss chard make great fall containers.

Winter

In winter, containers can be filled with boughs of evergreens. Some foliage plants, such as springerii, can be left to dry in the containers making a decorative display all winter. Hardy trailing plants including vinca and ivy can remain in the container all winter. Woody plants offer interesting textures in winter and broadleaf evergreens such as holly, daphnes, boxwood, ivy topiaries and small conifers offer interest all winter. Arrangements of red twig dogwood and evergreen branches make a delightful seasonal display in urns near entrances.

Containers with Foliage Plants

Foliage plants and woody plants will work best for containers and planters used as screens and space dividers.

CHOOSING AND COMBINING PLANTS

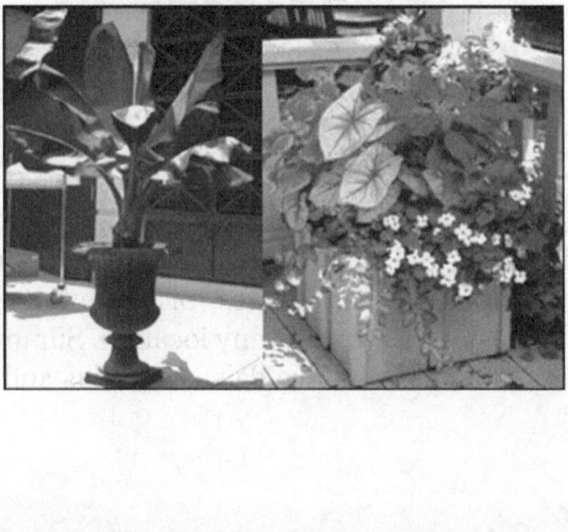

Think of a container garden as a living flower (and foliage) arrangement with roots! What look do you desire? Are you interested in something harmonious, dramatic or soothing? For a long period of enjoyment, keep the plants looking attractive by making sure the roots are healthy, and the foliage receives the light it needs to make its own food. Just one kind of plant in a container is the simplest to maintain. Plants that thrive in like soil, watering, and light conditions make successful combinations. When combining plants, size, texture, proportion, color, setting, and lighting all play a role.

MAKING HERB AND VEGETABLE CONTAINERS

Vegetable gardeners in urban areas are faced with lots of problems such as lack of space, soils contaminated with heavy metals such as lead and arsenic, shade from trees and buildings, and soil-borne diseases such as fusarium, pythium, rhizoctonia, and phytophthora. If you have a passion for vegetable gardening, consider growing vegetables in containers. Vegetables and even flowers and herbs grown in containers can be placed or moved to any spot such as windows, balconies, patios, and doorsteps where there is full sun.

The vegetables suitable for container gardening are the ones that require small spaces — particularly the dwarf or determinate types, that bear fruit or other harvestable parts over a longer period of time. They usually require full sun, more than six hours of sunlight a day — or partial shade, three to six hours of morning or early afternoon sunlight a day. Most fruit bearing vegetables such as tomatoes, peppers, cucumbers, squash, and eggplant require full sun. Leafy vegetables such as lettuce, cabbage, collards, mustard greens, spinach, and parsley can tolerate more shady location compared to the root vegetables such as turnips, beets, radishes, carrots, and onions. Most herbs can perform well in full sun and partial shade locations.

WATERING

Because containerized plants have a very limited soil volume, proper watering and fertilizing are critical in maintaining an attractive pot throughout the growing season.

How often you need to water a container varies depending on the time of the year, location of the container, how long the container has been planted, type of container, and type of plant material. There are no hard and fast rules when it comes to watering. That is why you have to be watching your containers on a regular basis and understand the requirements of the plants you choose to put in the containers. While there are all kinds of water meter devices on the market they are really no substitute for observing the plant and using your finger to see how moist or dry the soil really is.

Watering should be done when the plants need water and preferably before they start to show signs of wilting severely. Quality and overall heath of the plant are greatly affected by allowing the plant to wilt severely. Some plants can tolerate and prefer dry soils and if kept too moist will often decline. Succulents are an example versus tropicals. Understand the plant you have. Water it based on its needs and not some general rule of thumb.

When it comes to rules of thumb the best way to tell if a plant needs water is to feel the soil. And if the first inch or so of the soil is dry, water. Apply enough water each time so the whole soil ball becomes moist. This means using enough water each time so water starts to drip out of the drainage holes. This assures you have gotten the whole soil mass moist, and it also assures you have gotten some leaching of the soil.

Leaching means you move away any harmful build up of soluble salts in the soil with water. Soluble salts can come from fertilizer and water sources and leaching just prevents it from building up and causing plant damage. Damage can take the form of burned leaf margins and injured roots. Use well water or naturally collected water source and not a source of water that has been softened. Watering can be done using a hand held watering wand or if you have a lot of containers located in the same general area, drip irrigation systems can be set up to make watering more convenient and less time consuming.

Keep in mind that when using soilless potting mixes, they have a tendency to shrink away from the pot edges if allowed to dry out excessively. If this happens, rewetting becomes difficult because water tends to flow through the pot in the space between the pot and media leaving the media dry. If the container has dried out to this point, repeated watering will be needed to rehydrate the soil. If possible, immerse the container in a bucket or sink filled with water and leave it there allowing water to be absorbed through the holes in the bottom of the container. When you see moisture on the surface of the media, you know the soil is now moist.

To help retain moisture in the soil and to lengthen the interval between waterings slightly, you can incorporate one of the many gels available. This polymer based product looks like granular rock salt when dry and is mixed into the media before planting. Always read label directions for amounts to use based upon the size of the container and always mix the dry product. When these crystals are subjected to moisture, they expand taking on the consistency of gelatin. They now become reservoirs for water that a plant can use when the soil itself becomes dry. These products do not eliminate the need for timely watering. They just provide a slight cushion.

If you are growing perennials or evergreens in containers, be sure they are well watered before storing or covering them for the winter. Evergreens are especially vulnerable to winter desiccation if soils dry out during the winter.

FERTILIZING

Frequently watering containers results in the leaching of nutrients. These nutrients need to be replaced on a regular basis. A regular fertilizer program is needed to keep plants growing well and attractive all season.

There are several ways to supply nutrients. One is through the use of slow-release or timed release fertilizer. These products consist of water soluble fertilizer that is encased in a semi-permeable resin coating. When they come in contact with water, small amounts of nutrients are released to the soil for use by the plant. So, each time you water, the containers are getting "automatically" fertilized. These products are usually mixed with the potting media or placed on the soil surface at a rate specified on the label and based on the size of the container. Many of the products for use with annuals will supply nutrients for 3-4 months depending on the amount of moisture and temperature.

Fig. Selecting Fertilizer

Another way to fertilize containers is to use liquid fertilizers. These products are mixed with water according to label instructions and then applied during normal watering. It is suggested to apply fertilizer to the soil media when it is moist rather than when it is completely dry. This helps to avoid potential damage to the plant. Reapplying every two weeks or so will provide adequate nutrition for most containers. You can adjust the frequency of application based on overall appearance and growth. You can also fertilize with every watering using only a half strength fertilizer solution. Occasionally use just plain water to leach the pot.

The choice of fertilizer analysis will depend on the kinds of plants you are growing. High nitrogen sources would be good for plants grown for their foliage while flowering and vegetable crops would prefer lower nitrogen and higher phosphorous types.

GROOMING BASICS

Many gardeners are intimidated when it comes to training and pruning container garden plants. There comes a time when some grooming is needed to maintain an attractive container.

Grooming can consist of pinching, deadheading and general cleanup.

Pinching refers to removing the growing tip of a plant. When plants get too tall and start to get leggy, removing the top of the plant results in lowering the height as well as causing the plant to develop a bushier habit. Pinching also helps to maintain a certain shape and keep growth in check on some of the more vigorous plants. Cutting back vining plants would also fit into the category of pinching. To pinch, simply remove a small amount of the tip of a branch with either your fingers or pruning shears. Once pinched, new shots soon develop to fill in the plant. While pinching is not a daily task, look at the plant on occasion to decide if pinching will improve the shape and appearance.

Deadheading applies to flowering plants. After plants get done flowering, all that is left are spent flower heads. These spent heads are great places for disease to develop (botrytis) as well as cause the plant to bloom less. After the flower has started to decline and is no longer attractive, simply cut the flower stalk out. This cleans the plant up, reduces disease problems and encourages rebloom. Some plants don't require deadheading and are often termed "self-cleaning." Petunias, begonias and vinca fall into this category.

General clean up entails the occasional leaf removal due to insects, disease or just natural decline.

Keeping a container garden tidy is not difficult as it can be easily done when you walk by, or you can combine it with another task such as watering.

BUILDING SUCCESSFUL CONTAINERS

Building a successful container garden is no more complicated that choosing a container, filling it with media and planting the plants. Container gardens tend to need a bit more attention than a similar garden in the ground. You need to pay close attention to the potting media you use, how you water and fertilize and the types of plants you choose based on location of the garden. All of these will determine the success of the container.

First, fill the container with media. The section on selecting a soil media outlines the types of media available and what you need to consider based upon what you will be growing. In any case, all medias need to be well drained yet retain enough moisture for good plant growth. Bagged media have ingredient labels on them. If they are unclear, ask your retailer to explain what they mean and maybe even ask what type of media they prefer to use for their containers.

Bagged medias are easier to use when they are slightly moist prior to filling the container. Pre-moisten the media and then fill the container to within about an inch of the top of the pot. This space is referred to as head space and provides a place to put water and have it soak down into the container and not just run off the top.

What type of plants you choose will depend on location — sun or shade. How you arrange them depends on how you will be looking at the pot. If you will be looking at the pot from all sides the classic approach is to place tall plants in the center and then surround them with shorter mounding plants and finally finishing off with trailing plants along the edges. If it will be viewed from only the front locate the tall plants at the back and work forward with shorter plants and then trailing types.

If you are not sure of your design, set the plants on top of the media to visualize the arrangement before you start to plant. You can then adjust as needed. Once you have settled on a design, start planting. Remove the plants from their pots and look at the roots. If they appear to be coiled around the pot and seem pot bound, gently pull the roots apart to loosen them. Plant at the same level as they were grown in the original pot. If the soil ball seems a bit large for the spot you want to plant, gently squeeze the soil ball to help it fit.

How many plants you choose to use depends on how quick you want a "full" looking container. If you want a pot that looks finished right after you get done planting, you can space the plants very tight. If budget is an issue or you can wait a while, space plants out with room between them. In a few short weeks, the pot will have grown together with little if any empty spaces left.

After planting, water thoroughly. This allows the potting media to get in and around the root ball of the plants and eliminate air pockets. Make sure water drains out of the hole in the bottom of the pot. You may need to water several times to insure the media is thoroughly moist.

TROUBLESHOOTING CULTURAL DISEASE AND INSECT PROBLEMS

Aster Yellows (Phytoplasma)

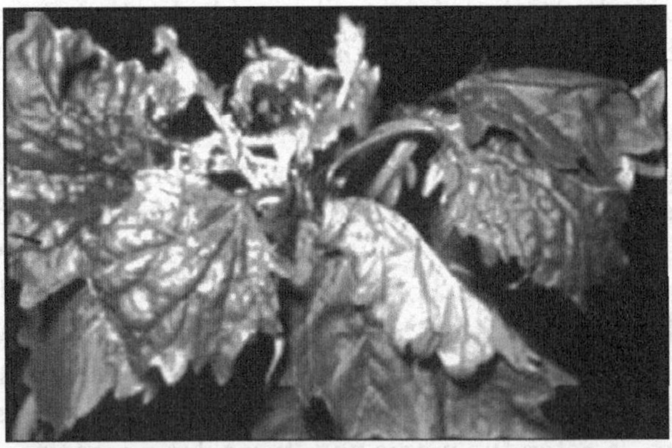

Symptoms & Life Cycle: Causes stunting, yellowish-green foliage and flowers followed by browning and death of the plant. Symptoms can vary from plant to plant and area to area due to different strains of aster yellow phytoplasmas. Phytoplasmas are bacterial-like but lack rigid walls. The organism overwinters in insects such as leaf hoppers. Disease passes into the plant during insect feeding.

Control: Destroy infected plants immediately to reduce the risk of additional insects picking up and transmitting the disease to other plants. Minimize insect wintering sites by mowing tall grass and reducing tall brushy areas.

Alternaria (Fungus)

Symptoms & Life Cycle: On calendulas, Alternaria causes small reddish to purplish spots. On other flowers the most common symptom is brown or dark brown spots on older leaves. Spots are circular to angular in shape. Spots can enlarge to ¼ to ½ inch in diameter and develop dark, concentric rings or ridges, causing a target-like appearance. Angular spots are caused by the veins stopping the expansion of the disease spots. Spotted leaves soon turn yellow, wither, and drop. Spots frequently covered with dark velvety conidia. Disease over-winters in plant residue left on or below soil surface as well as in the seeds. When weather is suitable in the spring, the fungus produces spores that are carried by wind and splashing rain. Repeat infections occur as long as the weather remains favorable (heavy dews, rainy conditions and temperatures between 66 and 73F.) for the disease.

Control: Rotate susceptible flowers with more resistant ones. Practice good sanitation. Since this fungus is a weak pathogen, keep plants healthy (proper fertility, watering, as well as limiting insect damage)

Anthracnose (Fungus)

Symptoms & Life Cycle: Anthracnose usually causes brown spots of various sizes depending on host plant and the anthracnose fungus causing the infection. Spots may remain small and limited in number or infect and kill the entire leaf. Anthracnose can also infect and kill plant stems. Symptoms will vary from plant to plant and will even vary on the same plant depending on pathogen, host, time of infection, weather and so on. The disease over-winters on dead plant material and spores can blow in the wind for many miles. Wet conditions promote the infection and spread of the disease.

Control: Grow plants in sun if possible. Provide plenty of good air circulation around individual plants. Fungicides and sanitation may help.

Gray Mold or Botrytis (Fungal)

Symptoms & Life Cycle: On tulips this disease is known as Tulip Fire. Gray mold appears as brown dead areas but with proper moisture, a gray fuzzy appearance occurs producing "spores." The pathogen can attack both flower and leaf buds, flowers and leaves, and stems and fruit. It can eventual kill many herbaceous plants. Several fungi cause Gray mold/botrytis. The organisms overwinter on dead and live plant material. It grows at temperatures between 32 and 84°F. It releases spores on rising as well as lowering humidity. However, it needs free standing water to infect. It is an air-borne disease and can blow for many miles.

Control: Sanitation and fungicides may help.

Powdery Mildew (Fungal)

Symptoms & Life Cycle: White powdery growth develops on upper and often lower leaf surfaces, young stems buds flowers, and even the fruit of some plants. Entire leaves and plant can be infected. Depending on the powdery mildew pathogen, the white growth may become mealy or felt-like and turn gray to tan.

Leaves may stunt, curl, become chlorotic and drop early. Flowers may distort. Some powdery mildew pathogens are very host specific while others have a broader range of host plants to infect. Mildew fungi can overwinter on dead plant material and flower buds. Spores are released in damp spring weather and are blown to uninfected plant tissue. Disease needs three or more consecutive days and nights where the days are warm and dry and the nights are cool and humid but plant tissue is NOT wet.

Control: Place plants in the proper growing site with good air circulation. Avoid shady humid areas. Follow good sanitation practices. This does not stop infection but can delay infection and reduce initial infection severity. Avoid over fertilization and watering. Grow resistant varieties. Fungus can mutate and attack what were resistant varieties whenever possible. If infection occurs early in growing season, consider a fungicide but if infection occurs late in the growing season, no fungicide treatment is recommended.

Root Rots (Fungal Water Molds)

Symptoms & Life Cycle: Pythium and Phytophthora are both water mold fungi that attach plant roots as well as the base (collar) of the plant stems when soil is kept too wet. Plants roots and stems rot and die. Top growth stunts, wilts, turns brown and dies. Infected tissue is brown to chocolate brown to black in color. Frequently tissue is "mushy." These pathogens spread by the movement of contaminated soil, tools, infected plants or splashing rain and wind.

Control: Make sure soil drains well and do not over water. Do not let pots/ containers sit in water filled saucers.

Sooty Mold (Fungal)

Symptoms & Life Cycle: Sooty mold fungi have a black mycelium and feed/ grow on insect honeydew. The honeydew is excreted by certain types of scales as well as aphids. The insects may not be on the plants that are covered by the sooty mold. The insects could be on the trees the containers are under. The severity of infection varies with the individual sooty mold pathogen as well as whether or not multiple sooty mold fungi are growing together. Sooty mold fungi continue to grow as long as there is adequate honeydew being developed. Sooty mold kills the foliage by reducing light to the chlorophyll which eventually dies and in turn cause the leaf to die.

Control: Eliminate or reduce the insects producing the honeydew.

Verticillium and Fusarium Wilt

Symptoms & Life Cycle: These pathogens live in the soil and attack plants that are under stress such as over watering, drought, over fertilization, and so on. Plants may be infected for awhile before showing symptoms. Initially plants wilt. Wilting tissue soon begins to yellow then turn brown and die. Internal discoloration or streaking of the vascular tissue occurs in most plants. The wilt diseases enter through the roots. They can survive in the soil as saprophytic (live on dead plant tissue) for years. Infected dead plant roots improve their survival in the soil. These wilt fungi are often moved with infected soil or infected plants.

Control: Avoid stressing the plant roots. Avoid overwatering, drought, over-fertilizing, deep cultivation, or use of herbicides. Buy resistant plants when possible. Destroy infected plants, and use sterilized containers and soil.

Aphids (Insect)

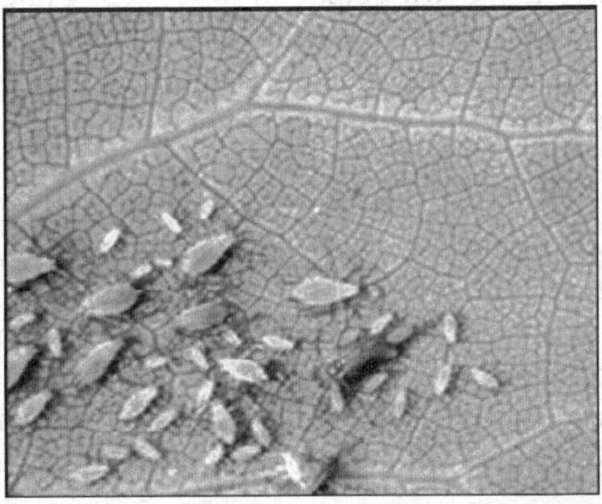

Symptoms & Life Cycle: Aphids have piercing, sucking mouth parts. On tender shoots and leaves, the feeding can cause distortion and puckering of plant tissue. The insects suck sap from plants. As the sap moves through the insect, necessary nutrients are removed and the excrement is called honeydew. Aphids overwinter as eggs. The eggs hatch in the spring. All newly hatched aphids are females who can produce live young in just a few days. When the aphids become overcrowded, the adults give birth to winged aphids that fly to other plants or other plant areas where they give birth to wingless aphids. As temperatures fall and days shorten, the aphids are stimulated to give birth to live males and egg laying females. The males only mate with egg laying females.

Control: There are insect predators such as lady beetles, lacewings, syrphid flies and parasitic wasps. There are also fungi that attack aphids. Plain water from a hose can wash them off. There are also both organic and inorganic insecticides. Follow all label directions and precautions when using any pesticide.

Fungus Gnats (Insect)

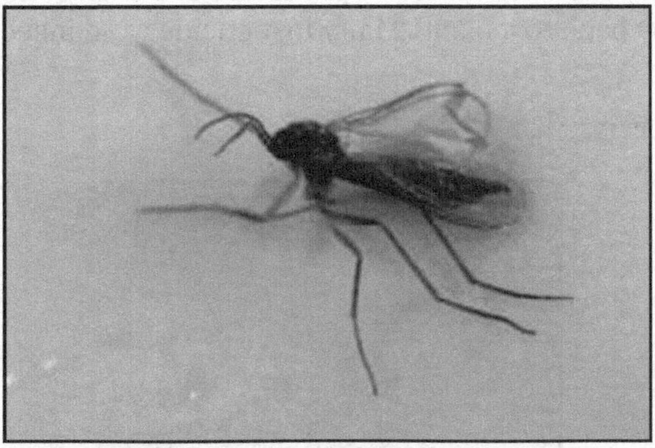

Symptoms & Life Cycle: Fungus gnats emerge from soil as adults. They are small, thin and dark in color. They can run across soil, plant and pots and will fly readily. They fly around people when disturbed. They are mostly a nuisance insect. Eggs hatch in potting soil as well as garden soil especially after the pot/container is potted and watered frequently to establish the plants. The prolonged moisture allows the insect time to complete its life cycle and emerge as an adult.

Control: No insecticide is needed. Carefully dry out soil without allowing the plant to wilt. This allows eggs and larvae in the soil to dry out and die.

Meadow Spittlebug (Insect)

Symptoms & Life Cycle: The immature insect produces protective foam that looks like "spit" around itself. This foam helps protect the insect from ultraviolet light, predators and insecticides. Adults overwinter, and lay eggs in the spring. Spittle-covered green nymphs become obvious on various plants during the spring. Later in the summer, the brown leafhopper-like adults are common.

Control: Use a strong spray of water to wash the foam and the insects off the plant. This expose the insects to predators and many will die before they can find their way back to a plant. Many insecticides and insecticidal soaps are effective too.

Mealybugs (Insect)

Symptoms & Life Cycle: Mealybugs typically are coated with a white waxy substance that covers the insect. When plants are grown in a very porous soil as often used in pots and containers, there are often mealybugs feeding on the roots as well as the above-ground portions of the plants. Feeding damage can cause distortion of leaves and death of plants especially small herbaceous plants grown outdoors.

These insects produce honeydew that is sticky. Feeding usually occurs on the underside of the leaf, where leaf and stem attach, or at the base of whorled leaves and on plant roots. When mealybugs congregate in masses, they cause the plant to look "cottony" in spots. Sometimes these "cottony masses" are egg clusters.

Control: Mealybugs are difficult to control especially if they are feeding on roots. The waxy coating protects them from insecticides. If only a few - try washing them off, or on a weekly basis, use a Q-tip to coat them with rubbing alcohol.

If plants are heavily infected - consider destroying the plants, if replanting, use new potting soil, or use suggested insecticides according to all label directions and precautions.

Whitefly (Insect)

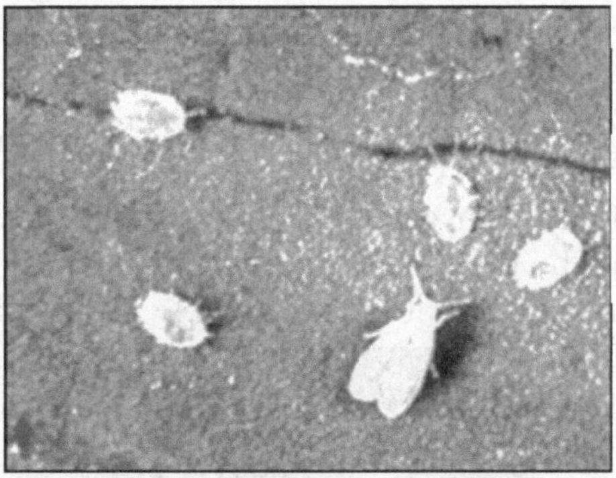

Symptoms & Life Cycle: Severe white fly infestations may cause mottling and yellowing of leaves as well as reductions in vegetable yields. Whiteflies are covered in a white waxy powder and adults are about the size of small "dandruff" particles. Whiteflies flutter around the plant when disturbed but soon return to the plant to continue feeding. Whiteflies found damaging cultivated plants in Illinois are not cold hardy this far north. They are shipped in as eggs on flowers and vegetable transplants grown in the South or they survive indoors on houseplants, herbs, and greenhouse crops. Eggs are very tiny and are on the upper leaf side. Adults as well as the tiny, clear oval nymphs tend to feed on the lower leaf surface.

Control: Use suggested insecticides according to all label directions and precautions. On vegetables make sure to follow pre-harvest interval. If there is no pre-harvest interval listed, do not use that insecticide. Find another that is labeled for controlling whitefly on vegetable.

Two-Spotted Spidermite (Insect-relative)

Symptoms & Life Cycle: The two-spotted spidermite is greenish to greenish-yellow with two dark spots on its back. It is also known as the red spidermite in Europe where it is red. But it is rarely red in the United States. Spidermites make single strands of silk, cause stippling (tiny yellow to whitish spots due to feeding of the foliage), and can cause the leaves to turn brown and die when infestations are severe. The webbing is frequently washed away by rains and blown away by the wind. The females overwinter in leaf litter or under the bark of several different types of trees and shrubs. Reproduction can occur throughout the growing season as long as the weather is warm enough. From egg to adult takes a short time, so in Illinois twenty or more generations per growing season is common.

Control: Beneficial; mites may feed on them. Predator mites usually crush red and two-spotted spidermite and other harmful spidermites tend to crush green. Look for the webbing or hold a white sheet of paper under the plant and rap the plant vigorously on the paper. Watch closely for tiny mites crawling while the debris will not move. Use a miticide, insecticidal soap or summer spray oil only if the moving mites are the "bad" mites. Follow all label directions and precautions.

CONSTRUCTING AND CARING FOR CONTAINER WATER GARDENS

A decorative pot of water garden plants on your patio or deck is not only beautiful, it might just be the lowest maintenance gardening you will ever do. You can't over water them and it is very hard to under water them, too.

From one gallon to 20 or even to 500 gallons, any tub, pot, barrel, or container that can hold water can be used. The larger it is, the easier it will be to maintain healthy fish, if you desire them. Pumps and fountains are not required, but do add to the pleasure of a patio water garden and can help in maintaining fish.

Since water weighs eight and a third pounds per gallon, a 20 gallon container could weigh over 200 pounds when the water, plants, soil, and container are all added up. Select a location for the water garden that can hold the weight. A tall deck may not be able to hold a large container. If possible, place it where it can be seen from the indoors as well as the patio. If you add a fountain, the water garden container will become a focal point for the patio. Fountains do need electricity, so locations near outlets help shorten the length of necessary cords.

You will often see recommendations that water garden containers be placed in full sun conditions. If you don't have full sun, don't worry, there are plenty of aquatic plants that will do fine in a shaded container water garden, although many water lilies bloom better with more hours of sunshine than less. Full sun conditions can help to promote algae and warm water temperatures too high for goldfish. Three to five hours of direct sunlight is all that is necessary for most aquatic plants and protection from afternoon sun is often beneficial when they are grown in a pot.

Your water garden location should have easy access to the water supply. Containers with fountains will lose water more rapidly than ones without them. Keep in mind that water splashing out of fountain containers may allow algae to grow on paving surfaces, which could then become slippery.

CONSTRUCTING AND CARING FOR CONTAINER WATER GARDENS

THE CONTAINER

Picking the best container for a patio water garden depends what kind of pot would match your garden decor and what kind of plants and fish you want to add. The requirement that it has to be water tight is only partially true. A small leak means you will have to add water more often, which can be a good maintenance tool as it keeps the water fresh. Plastic liners can be added to wooden half-barrels and waterproof coatings can be painted or sprayed into other pots to make them hold water.

It doesn't have to look like a flower pot to be used as a water garden container. Old bath tubs, galvanized wash tubs, horse water troughs, kid swimming pools, old rubber boots, and other household items have all been turned into water gardens. Use your imagination.

If you want to use a terra cotta or ceramic flower pot that already has a drain hole, you will need to try to plug it with a rubber or cork stopper. Try coating the plug with silicone sealant to help fill the gaps and to create a good watertight seal.

Wooden half barrels can be bought fresh or they may have held whiskey or wine depending on the region of the country you are in. Sometimes they have been charred on the inside to aid in the filtration of the liquid contents. It is easy to drop a pre-formed plastic liner into the wooden barrel to ensure it is water tight and free from impurities. Some hard plastic liners do not fit all barrels and they can collapse into the barrel and crack, when the weight of the water is added to it. Be sure to get a properly fitted plastic liner that sits flat on the bottom before the water is added.

Fill a fresh wooden barrel with water and see if it leaks. Small leaks will often plug themselves, as water saturates the wood and it swells to seal the cracks. Painting the interior of a barrel can aid in waterproofing. Use marine grade paint. If you purchase an old used barrel, make sure the metal hoops still hold the wood slats in place and that the barrel sits level.

Inexpensive plastic sheeting can be used to line a barrel that leaks too much, but it usually must be replaced every season and can be easily damaged when moving the pump or plants around. Lay the sheet into the barrel and add water, smoothing out wrinkles as you go. When the water is near the top, you can staple the plastic to the wood, then trim off the excess plastic. Heavy duty pond liner can be a longer lasting alternative to plastic sheeting because it is designed to hold water and will not decay in the ultraviolet light from the sun.

Darker-colored interiors seem to work better on water garden containers, because they reflect less light back into the water and visually appear to be deeper. Less reflected light means less light is available for algae to grow. Any algae that does grow is harder to see on the darker colored surfaces. Darker pots absorb more heat, which is good in the spring to speed plant growth and may be bad in the summer if it heats the water too much for fish.

Aquatic plants can be installed directly into smaller containers, while larger containers can hold several plants still in their nursery pots. The only rule for pot size is to make it large enough to look proportionate with the plants that are going in it. If tall plants are being used, consider the wind and make sure the pot won't be tipped over. Water garden containers can be as small as table top or the size of huge caldrons. The pot can even be filled with decorative gravel. The container does not have to be visible. It can be buried in the ground or hidden behind decking. If it is buried in the ground, it can have a fountain or waterfall pump circulating the water to create more interest as water shoots up from the gravel or cascades down the rocks. This type of installation creates a very natural bird bath.

PLANTS

Aquatic plants grow in a variety of conditions. Some grow completely underwater, some float around on the surface, some root on the bottom while their leaves float on the surface and some aquatic plants grow on the shoreline with their roots in muddy water.

Shoreline Plants

Plants that grow entirely under water are sometimes called oxygenators. They do supply some oxygen to the water during photosynthesis, but they also consume oxygen 24 hours a day through the process of respiration.

They are not necessary in a container water garden, but they can be used if desired, especially if an ecosystem approach is being used for maintenance, where each component of the system is being used. They can be beneficial as they remove nutrients from the water that would otherwise feed algae.

A few of these plants grow mostly under water but often have above water stems that can be very attractive in containers, such as the plant called Parrot Feather. Wild Celery (Vallisneria sp.) flat ribbon-like, translucent, pale-green leaves. It can tolerate shade, part-shade, and sun in water that is 6 to 24 inches deep.

Fanwort (Cabomba canadensis) Bright green feathery foliage. It can grow in sun to part shade in water that is six to 12 inches deep. Anacharis (Egeria densa) Dark green leaves, and it sometimes blooms with tiny white flowers that float on the surface. It can grow in water one foot to over five feet deep. Parrot Feather (Myriophyllum aquaticum) Grows in shallow water, partly submerged and partly floating. Can grow three to six inches out of the water, or it will hang several feet over the edge of water garden containers. Grows in full sun or shade.

Floaters

Floating plants are great additions to many water gardens. They have root systems that hang down into the water. They filter the water of nutrients that would otherwise help algae grow. They shade the water, sheltering fish and preventing sunlight from benefiting algae. Some floaters such as Water Hyacinth have pretty flowers too. Many floaters, including Salvinia and Water Hyacinths, are banned from several states because they can spread rapidly in natural waterways.

Giant Velvet Leaf (Salvinia longifolia) A small, tropical, floating fern with pale green, round, hairy leaves. Water Hyacinth (Eichhornia crassipes) A common plant that cannot tolerate frost. Long, trailing roots, balloon shaped stems, and spikes of pale lavender flowers. Water Lettuce (Pistia stratiotes) Velvet-covered green leaves forming a rosette of foliage that looks like a floating head of lettuce.

Water Lilies and Lotus

These plants have their root systems firmly anchored in the bottom mud, while the leaves and flowers either float on the water's surface or are held above the water. There are several dwarf varieties of both lilies and lotus that are suitable in container water gardens. They all need full sun for best blooming, but some can tolerate a lot of shade, if blooms are not the primary focus.

Hardy water lilies live where the water's surface freezes, but the roots don't freeze down deep in the pond. They only bloom in the daytime and are usually not fragrant. The leaves have smooth edges.

Tropical water lilies cannot survive freezing. They are often fragrant and the flowers come in blue and purple shades that are not found in hardy water lilies. Some even bloom at night. The leaves have scalloped edges and are often mottled in red tones. They require warm water, at least a half day of full sun, and around a foot of water over the crown.

Froebeli - A hardy, red lily that can stand the heat that might build up in a small container.

Joanne Pring - A hardy, miniature, pink water lily with green leaves.

Tetragona - A hardy, miniature, white water lily with red mottling on the leaves.

Helvola - A hardy, miniature, yellow water lily with red mottling on the leaves.

Yellow Pigmy - A hardy miniature that is a prolific and consistent bloomer.

Hilary - A tropical, day blooming, pink water lily with green leaves.

Red Flare - A tropical, night blooming, red water lily with maroon leaves.

Momo Botan Lotus - Grows to a height of 24" with rose blooms July - September.

Wan-er Hong Lotus - Grows to a height of 12" with white blooms June - September.

Shoreline Plants

Depending on the species, some of these plants like to always have several inches of water over the crown of the plant, while others can easily tolerate merely damp garden soil around the roots. Planting shoreline plants in a container water garden should take into account how much water over the root system the plant prefers. Arrowhead (Sagittaria sp.) Green arrowhead shaped leaves. White, one-inch flowers five or ten to a stalk. Grows from 12-48 inches tall, in sun to part shade, in streams and along shores in shallow water.

Cattail

Creeping Jenny (Lysimachia nummularia). Grown for its ground-cover-like creeping runners that hang over the sides of containers. Enjoys shallow water conditions and tolerates bright, indirect light or partial shade.

Blue Flag Iris (Iris versicolor) Blue flowers in late spring with flat foliage. Grows from 24-30 inches tall and will tolerate full sun to part shade. Tolerates several inches of water over the crown to just moist garden soil. Houttuynia "Chameleon" (Houttuynia cordata) Multicolored leaves of red, cream, pink, and green. Perennial creeping plant that grows to six inches tall. Grows well in part shade and hangs over the sides of water garden containers.

Lobelia

Dwarf Papyrus (Cyperus isocladus) Green grass-like foliage in an umbrella shape on top of a stalk. Flowers amidst the leaves are not showy. Grows 12-18 inches tall and tolerates sun to part shade. Needs to stay wet and will not tolerate cold weather. Water Canna Many flower colors are borne on top of three to five foot tall stalks of wide green leaves. Some have red or multi-colored leaves. Grows in moist soil to three inches of water.

Taro Huge green leaves up to seven feet tall. Tropical that tolerates shade. Grows in damp soil to submerged several inches. Corkscrew Rush (Juncus effusus) Interesting twisted and curled wiry leaf. Grows to 24" tall and tolerates full sun to part shade. Prefers one to three inches of water over the crown.

PLANTING THE WATER GARDEN CONTAINER

For healthy plant growth, each water garden container must be able to hold enough water to cover each plant to its preferred depth. For example, let's take an aquatic canna growing in a six inch tall nursery pot that prefers less than three inches of water over the crown. If it is set on the bottom of a 24 inch deep container, the plant will drown. If the same canna pot is set on some bricks or upside down pots so that it is raised up and the crown of the plant is near the water's surface, the plant will thrive.

The term water garden container encompasses the small pot that holds a single plant to a large one that holds many plants, fish, and fountains. The single plant pot is easy to create and maintain. If it holds an underwater plant or a floater, just fill it with water, add the plants, and you are done.

If it holds a shoreline plant, add some heavy clay soil, up to an inch of gravel on top to keep the mud in place, add the plant and water, and you are done. Don't use potting soil as many of the ingredients are too light and they will float, causing the water to always be a mess. If you don't have any clay soil, you can use generic kitty litter - it is made from clay. Do not use any that have deodorizing crystals or other ingredients.

Many aquatic plants look good all by themselves in a single pot. Dwarf lotus and water lilies can be planted in table top bowls. Larger specimens look nice in large oriental ceramic vases.

Larger pots can be planted just like the small one with several types of plants all growing together in the container. Large containers can also hold smaller pots to make a water garden arrangement that is easily changed as the plants grow or as the gardener decides to rearrange it. If the plants used prefer different depths of water, some can be set on pedestals to raise them to the proper level. Bricks, stones, and flower pots can all be used. Solid pedestal materials like bricks replace water volume, so if fish are being added, they will have less water available. Plastic pots will float and be difficult to use. Upside down clay pots often work very well.

A mix of container sizes and planting configurations can be used to make a flower pot arrangement on the patio. Just like with other patio pot arrangements use various sizes, shapes, and colors of containers to make your own personal style. A mixed aquatic container can have the same style as many terrestrial potted arrangements; for example, a tall, spiky plant in the center or in the background, surrounded by showy flowers or brightly colored leaves, and as a final touch, plants that cascade over the edge of the container. Many aquatic plants have brightly colored foliage. Use them side by side for contrast and to attract attention.

Tall, spiky plants include cattail, yellow flag iris (Iris pseudacorus), sweet flag(Acorus calamus), rushes, papyrus, and canna. Broad-leaved plants include lotus, taro, arrowhead (Sagittaria), calla lily, and lizard tail. Flowering plants include water lilies and water bluebell. Cascading plants include parrot feather (Myriophyllum aquaticum), aquatic mint (Mentha aquatica), houtuynia, and water celery. If the container has enough surface water showing, then the addition of floating plants is always welcome. Add water lettuce, Pistia stratiotes, or water hyacinth, Eichhornia crassipes.

Many plants fit into more than one category. Don't crowd too many plants into a container. Aquatic plants are often very fast growers. If there is too much fertilizer in the water, the floaters will grow rapidly. It is easy to take a few out from time to time and toss them under some shrubs for fertilizer or add them to the compost pile.

FISH AND OTHER ANIMALS

Several kinds of aquatic animals can be added to water garden containers. None are required, but they can help create a balanced ecosystem. Goldfish are the first animal most people think of, but many small aquarium fish can live in an appropriate water garden container. Koi are a type of carp, and they grow over two feet long, so they won't do well in anything but a large container. Depending on the kind of fish and the amount of maintenance provided, the general rule for small fish is at least a gallon of water for each one is necessary if filters and pumps are used, and it is better if there is only one fish for every five to ten gallons if little maintenance will be available.

Just like in an aquarium, water quality must be maintained for good fish health. Chlorine and chloramine must be removed, either by waiting until they dissipate from the water or by adding an anti-chlorine product. Toxic waste products from fish are broken down by a variety of bacteria. Bacteria are required for fish health and do not come to a water garden on their own. They can be added from products purchased at the pet or water garden supply store. Fish will need to be fed except in large containers that have been established for such a long time that algae and other food sources have developed and are being maintained. Snails can be added to water garden containers to help control algae and to consume organic matter. Frogs will tend to hop away from any container that they can get out of. Toads are not aquatic, except during breeding season and won't stay in a water garden container.

PUMPS AND FOUNTAINS

The sound of moving water is very relaxing to many people. Small pump kits and fountains are readily available. Do not set the pump on the bottom of container where it will suck in too much debris and get clogged. As the pump agitates the water's surface, it reduces the attraction to mosquitoes, but it doesn't prevent them entirely. The pump can be connected to a timer that turns it off at night when no one is around. Mosquitoes can lay their eggs at night when the pump is off.

PROBLEMS - ALGAE AND MOSQUITOES

Algae

Algae are weeds in the water garden container. They can be controlled through natural means and through chemicals. Algae need water, sunlight, and nutrients to grow. We can't take away the water, but we can limit the sunlight and nutrient levels. Shade can be created by locating the pot where it doesn't get too much sun and by planting lots of plants in the water garden. Excessive nutrient levels in small containers tend to come from the fertilizer tablets placed in the soil of the plant pots. Make sure such tablets are completely covered with soil.

Underwater plants and floating plants get their nutrients directly from the water and compete directly with algae for nutrients. Floating plants shade the water, too. Potted plants get the majority of their nutrients from the soil in the pots and only compete for sunlight.

There are a few algicides that are safe for other aquatic plants, fish, and the other organisms found in backyard ponds and water garden containers.

Algaefix from Aquarium Pharmaceuticals was the first of these products and can be used in aquariums as well.

Mosquitoes

Mosquitoes only become a problem in water garden containers that are not maintained. If small aquarium fish or goldfish are in the container, they will eat any mosquito larvae. A natural product called Mosquito Dunks contains the bacteriaBacillus thuringiensis subspecies israelensis. It can be purchased inexpensively to completely eliminate mosquito larvae in all landscape situations that can hold water. The product can be placed in gutters, rain barrels, swimming pools, birdbaths, animal watering troughs, and anywhere else water accumulates. This product is effective, long lasting, and safe for people, pets, and fish.

Attempting to overflow the water garden container to get rid of mosquito larvae can be partially effective. Some larvae escape by swimming to the bottom of the container when the water starts to flow. Flooding the container trying to get rid of every last larvae can waste a lot of water.

Adding a small amount of vegetable oil on the water surface can temporarily prevent the larvae from surviving. It can be unsightly to see an oily surface and it coats plants and pots. It doesn't work if there are fish or fountains in the container. Water garden containers can be filled with gravel after the plants and water have been added. The water will not be visible or available to mosquitoes.

SEASONAL MAINTENANCE

Spring, summer, and fall maintenance is easy for water garden containers. Simply add water to keep it at the proper level and feed the plants and fish as necessary. Occasionally, the water may turn tea-colored though the decay of leaves or the pot may fill with unsightly debris. Empty the container, rinse it clean, and refill as quickly as possible to prevent the plants from drying out. Do not scrub the container clean as this will remove too many beneficial bacteria growing on the container's surfaces.

Winter maintenance depends on the type of plants, type of container, and the climate of the container's location.

There are three basic overwintering styles among water garden plants. First, there are aquatic plants that grow in cold northern climates and can easily handle being frozen solid. For example, think of cattails growing in a road side ditch in Minnesota. The second group is tropical aquatic plants that are never exposed to frost, and many won't even tolerate water temperatures dipping into the forties. These are plants like those that grow in the Everglades or the Amazon. In between these two extremes, the third group is aquatic plants that can tolerate cold water but not being frozen. Many of these plants have leaves and stems that die when the water gets cold, but the roots overwinter as long as they do not freeze. Hardy water lilies and lotus are good examples of this type of over wintering style.

If you live in a cold climate and have cold-tolerant plants, you can leave them outdoors as long as the container can tolerate being frozen. The tropical plants can be taken indoors and used as house-plants or they can be allowed to die with the cold weather. Some tropical plants require a lot of sun, so if there isn't enough sunlight coming in through the windows, they may not survive the winter anyway. Many of the third group of plants actually need a cold-induced dormancy period. If taken indoors to protect them from the frost, they will slowly die, and if left to freeze they will also die. If their pot can be placed at the bottom of a pond where they can have cold water without freezing they will survive the winter just fine.

If you live in a warm climate, you should stop fertilizing the plants for a few weeks in the fall to allow them to stop blooming as the day length decreases. When the plants begin to show active growth, they can be fertilized again. If it is not possible to bring in the entire container of cold weather plants, it may be possible to just bring in the smaller nursery pots and place them in plastic bags or tubs with a small amount of water in a cool, dark basement until spring.

Another cold climate possibility is to use a bird bath or water garden pond de-icer to keep the container free of ice. If possible, move the container close to a building wall and wrap insulation around and over it. Deicers are not designed to keep the water warm - they just keep the water above freezing. Tropical aquarium fish should be brought indoors long before the water gets cool. Goldfish can tolerate cold water, but they too should be brought indoors before the water temperature drops too much. If fish are placed in new aquariums, proper filtration will be required.

MAKING HANGING BASKETS

Hanging baskets are container gardens that are suspended overhead. They are great for giving the illusion of bringing down the height of very tall spaces and making great visual impact. Planting hanging baskets is not much different than planting containers that are on the ground.

As far as a choice of container there are basically three: solid plastic, wood and wire. Plastic and wood are very easily handled by filling with media and planting. With wire, you need to first line the container with something to keep the soil from falling out. This is done with either sphagnum moss or pre-formed coir basket liners. If using sphagnum moss, wet the most thoroughly before using. Then, taking handfuls off moss, pack it into the wire basket from the inside of the basket. Do this until the whole frame has been lined forming a pocket in the center that is then filled with media.

There are also pre-formed liners made from coir (coconut fiber). These forms come in various sizes and are placed over the top of the container and gently pushed down so the liner conforms to the container. With lined wire baskets, watering and keeping the container moist is of major concern. These types of containers loose a tremendous amount of moisture through the sides as well as through normal drainage out of the bottom. To reduce the moisture loss and cut down on watering frequency, it sometimes helps to incorporate a barrier to reduce moisture loss. Simply lining the inside of the moss lined basket with plastic prior to filling with media will help reduce moisture loss. Be sure to punch a few holes in the bottom of the liner for drainage. There are also coir liners available that have a plastic liner sandwiched between the coir layers to help reduce moisture loss.

After lining, fill the container with media. Most hanging baskets are planted at the top. If you are using moss or coir fiber lined baskets, you also have the opportunity to plant through the sides. If baskets will be hung at or slightly below eye level, follow a traditional planting scheme for ground containers. Locate upright or taller plants in the center with shorter and trailing material towards the middle and edge. If the basket will be located well above eye level, the tall center plants can be eliminated and use mainly rounded or trailing forms of plants. Baskets planted only with trailing types of plants may take a while to achieve a "full" look but soon will hide the basket completely. Some trailing types to consider are the trailing petunia, bacopa, scavola, sweet potato, verbena, calibrachoa, dichondra, ivy and lysimachia.

Planting through the sides has its advantages if you want a completely full looking basket and only want to use non-trailing types of plants. Plants such as wax begonia, pansies, alyssum, brachyscome or impatiens can be used. To plant into the side walls, the suggested method is after you have lined the basket with moss or inserted the liner, cut holes into the sides of the liner spaced around the basket starting at the bottom. Fill with media into the bottom of the hole. Insert the plants though the hole and fill in with soil to the level where the next series of holes will be cut. Insert plants and finish off with media, planting the top last.

After planting, water thoroughly to get media around the roots of the plants. Also keep the moss moist to help the plants to root quicker.

One of the major maintenance issues with baskets is watering. Both solid baskets and especially moss lined baskets dry out quickly. This is especially true later in the season.

When watering, fill the top of the container with water and allow it to drip out of the bottom of the basket each time you apply water. Try to maintain uniform soil moisture.

Feel the soil or watch the plants for signs of slight wilting. Frequent severe wilting will impact upon the quality of the basket. With frequent watering, fertilizer application becomes important. Apply fertilizer every two weeks or use one of the slow-release types of fertilizer mentioned in the fertilizer section. Combining slow release with a half strength liquid fertilizer program will insure good looking baskets through the season.

LANDSCAPING WITH CONTAINERS

While an individual well planted container can be quite impressive, think what might happen if you put several containers together in a group.

Putting containers together that have different heights, textures and styles can have great impact. A group of containers also offers you great flexibility to move them around and rearrange them for different looks. It also offers the chance to take a container "out of service" if it starts to look bad. Generally, this does not affect the overall display.

Putting containers together as a grouping can also have its advantages especially when they are first planted. Newly planted containers often don't have that "full" look right away. By grouping containers closer together, you can give the illusion of fullness. As the containers fill out, they can be spread further apart to present a totally different look. The same number of containers now takes up more square feet and looks like you actually added containers to the display.

Container height can also be varied simply by sitting some containers on overturned pots, bricks or other things to give some elevation.

Placing container groups by the entrance to the home helps to draw attention to the door and act as a sign of welcome. Grouping containers on a porch, balcony or patio has the affect of softening what might be a very stark, open space. It can make the place a refuge from all else that is happening in the world.

While most containers are found as a stand alone element, think about incorporating them into the garden itself. Large, bold containers worked into a flower or shrub border now becomes part of the planting bed. They are looked at as specimen or sculptural accents. The base of the pot is surrounded by annuals or perennials, the container is a piece of art, and it is topped off with a floral display.

Landscaping with containers can be as simple as one well-chosen and planted pot to a group of several pots making a statement and colorizing an otherwise unplantable area.

GROWING VINES IN CONTAINERS

Vines in containers offer great vertical accent. They can also be treated as specimen plants. There are any number of easily grown annual vines that offer not only flowers but interesting foliage as well. Some to consider include; morning glory, thunbergia, cardinal climber, cypress vine, sweet pea, hyacinth bean, mina lobata, snail flower and love-in-a- puff.

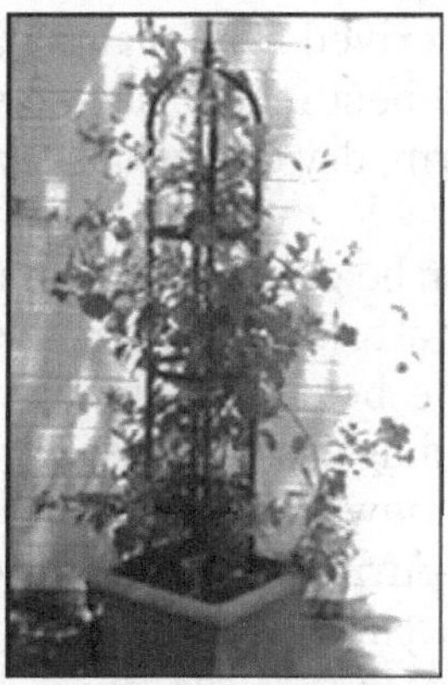

Most of the annual vines support themselves by either twining or using tendrils. Because of this some type of support is needed. The support you choose can be an elegant obelisk or as simple as bamboo stakes or even sticks from trees and shrubs. Supports with rough textured surfaces often offer better "gripping surfaces" for the vine as opposed to smooth slick metal rods.

Whatever you choose as a support make sure to insert the support in the container before you plant the vine so that it is there and waiting for the vine to climb on. Choose a support based on the vigor of the plant. Many container vines quickly out grow their support because the garner didn't realize how big the vine would get. Also make sure the support is firmly anchored. As the vine gets larger and heavier, it also becomes more prone to being knocked over during windy days.

Vines can also be used in hanging baskets. Not only will they grow up the wires supporting the basket, but will also hang gracefully down the side much like a curtain.

HOUSEPLANT

Houseplant, any plant adapted for growing indoors. The most common are exotic plants native to warm, frost-free parts of the world that can be grown indoors in colder climates in portable containers or miniature gardens. Most houseplants are, therefore, derived from plants native to the tropics and near tropics. Those that make the best indoor subjects are the species that adjust comfortably to the rather warm, dry conditions that generally prevail in indoor living spaces.

Although many plants can be grown successfully indoors, there are certain groups that, because of their attractiveness and relative ease of maintenance, are generally considered the best houseplants. These include the aroids, bromeliads, succulents (including cacti), ferns, begonias, and palms, all of which have long been favourites. Somewhat more demanding are those that are grown primarily for their flowers—African violets, camellias, gardenias, geraniums (Pelargonium species), and orchids.

HISTORICAL BACKGROUND

Paintings and sculptures make clear that the practice of indoor gardening can be traced at least to the early Greeks and Romans, who grew plants in pots and perhaps brought them into their homes. The older civilizations of Egypt, India, and China also made use of pot plants but usually in outdoor situations, often in courtyards that were extensions of the house; and for centuries the

Japanese have carried on the dwarfing of trees and other plants for room ornaments. But the popular art of growing houseplants did not receive much comment until the 17th century, when, in The Garden of Eden (1652), Sir Hugh Platt, an English agricultural authority, wrote of the possibility of cultivating plants indoors. Shortly thereafter, glasshouses (greenhouses) and conservatories, which had been used during Roman times to force plants to flower, were built in England and elsewhere to house exotic plants. In mid-19th-century England and France, books began to appear on the growing of plants in private residences, and the use of enclosed glass cases of plants (the wardian cases, or terraria) became popular.

TYPES OF HOUSEPLANTS

There are thousands of tropical and subtropical plants that can adapt to growing indoors. Although some fancy exotic species do well only in a humid conservatory or a glass-enclosed terrarium, a great many species have been introduced that endure the adverse conditions of dry heat and low light intensity that prevail in many houses. A selection of the more widely favoured houseplants follows, under two sections: foliage plants, some of which also bear interesting flowers; and flowering plants, species kept primarily for their flowers.

Foliage plants

In the aroid family, which has provided a range of long-lived houseplants, most prominent are the philodendrons. These are handsome tropical American plants, generally climbers, with attractive leathery leaves, heart-shaped, and often cut into lobes. Monstera deliciosa, or Philodendron pertusum, the Swiss cheese plant, has showy, glossy, perforated leaves slashed to the margins.

The dumb canes, of the genus Dieffenbachia, appear in a number of attractive species. They are handsome tropical foliage plants usually with variegated leaves; they tolerate neglect and thrive even in dry rooms. The Chinese evergreens, of the genus Aglaonema, are fleshy tropical Asian herbs of slow growth, with leathery leaves often bearing silvery or colourful patterns; they are durable and are tolerant of indoor conditions. Members of Scindapsus, popularly known as pothos, or ivy-arums, are tropical climbers from the Malaysian monsoon area; their variegated leaves are usually small in the juvenile stage. They do well in warm and even overheated rooms. The peace lilies (not a true lily), of the genus Spathiphylla, are easy-growing, vigorous tropical herbs forming clumps; they have green foliage and a succession of flowerlike leaves (spathes), usually white. Species of Anthurium, many of which, such as the flamingo flower, have colourful spathes, do best in humid conditions. Caladium's tropical American tuberous herbs produce fragile-looking but colourful foliage; they keep surprisingly well if protected from chills and wintry drafts.

Begonias, with their often very decorative leaves, have long been favourites among houseplants, but, with few exceptions, they require more humidity and fresh air than the modern home provides. Begonia metallica, with its olive-green, silver-haired foliage; B. masoniana, with beautiful green, puckered leaves splotched brown; and B. serratipetala, with small leaves spotted pink, are examples of types more resistant to dry rooms.

There are many small foliage plants, often with strikingly patterned foliage, native to the tropical forest floor, some of which have become remarkably good houseplants. Among them are several prayer plants (Maranta species), which fold their attractive leaves at night; and the exquisite Calathea makoyana, or peacock plant, with translucent foliage marked with a feathery peacock design. Pilea cadierei, or aluminum plant, is easy to grow; it has fleshy leaves splashed with silver. Codiaeum species, or crotons, are multicoloured foliage plants that need maximum light and warmth to hold their leaves and coloration well. Although primarily thought of as bedding plants, the varicoloured coleuses, or painted nettles, can decorate a sunny window with a brilliant array of leaf patterns. Peperomia species form miniature rosettes or vines with waxy foliage, corrugated and decorated either with silver or creamy white.

Bromeliads constitute a plant family peculiar to the Western Hemisphere; they dwell on trees and rocks (as epiphytic plants) or on the forest floor (as terrestrial plants) and usually form rosettes of leathery, concave leaves, many with bizarre designs or striking variegations. Their flowers may be hidden deep in the centre of the rosette, surrounded by a cup of brilliant crimson inner leaves, as in Neoregelia and Nidularium. Species of Aechmea and Guzmania form colourful spikes or heads of long-lasting leathery bracts or bright berries. Billbergia species are tubular in shape; their showy flower stalk, with blue flowers, is often pendant. Most forms of Tillandsia and Vriesea have spear-shaped, flattened, colourful flower spikes. The earth stars of the terrestrial genusCryptanthus are more or less flattened rosettes with striking leaf design, mottled, striped, or tiger-banded in silver over greens and bronzes.

Succulents

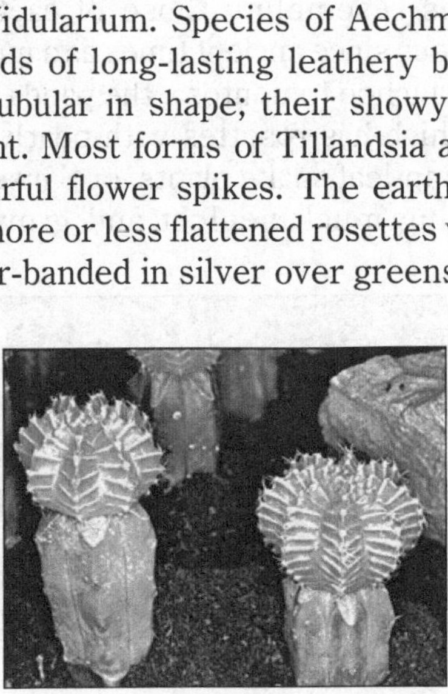

Cacti, most members of which are native to the Western Hemisphere, have developed a special capacity to store water in thick, fleshy bodies. They thrive in much sunlight and need very little water. There are many often curious forms: the tiny button cactus, Epithelantha; the myriad pincushion species of Mammillaria; Parodia, or Tom Thumb cactus; andRebutia, the pygmy cactus. The last two bloom when young and tiny. Other forms include Gymnocalycium, or chin cactus; Notocactus, or ball cactus; Echinocactus, known as barrel cactus; various Opuntia species, including bunny ears and chollas; and Cephalocereus, or old-man cactus, with its glistening white hair. Larger cacti includeCereus and its relatives, often night-blooming, and the giants of the desert, such as the saguaro (classified as Cereus giganteus or as Carnegiea gigantea), with branching columns up to 50 feet (15 metres) in height. Cacti of tropical forests include the epiphytic Rhipsalis, found also in Africa, Madagascar, and Sri Lanka, and the near-epiphytic leaf, or orchid, cacti, Epiphyllum, which bloom in many colours.

Succulents other than cacti have also contributed favourite subjects for indoor growing. A typical stem succulent is Euphorbia, with its often angled candelabra-like columns resembling those of cacti. Leaf succulents are represented by Aloe, famous since ancient times as a medicinal plant; Echeveria, or hen and chickens; Kalanchoe tomentosa, the panda plant; Crassula, the jade plant; and Haworthia, which has rosettes with pearly dotted leaves. Durable pot plants include the strap-leaf snake plants, or Sansevieria species; they are remarkable for tolerating much neglect and growing in less than ideal locations.Trees

Dracaena, the dragon trees, includes such houseplants as D. marginata, from Madagascar, which forms clusters of twisted stems topped by rosettes of narrow, leathery leaves. Other examples are D. deremensis'Warneckei,' with its handsome, symmetrical rosette of sword-shaped, milky-green leaves with white stripes; and D. sanderiana, the ribbon plant, a diminutive and slender, highly variegated species that can be grown in water. Similar in appearance is Pandanus veitchii, which has a rosette of leathery, sword-shaped leaves—glossy green and banded white—arranged in spirals.

Several subtropical evergreens can be grown in cooler locations indoors. Preeminent among them is the Norfolk Island pine (Araucaria heterophylla, or A. excelsa)—not a true pine—an undemanding graceful conifer with tiered branches of fresh green needles; it is long-lived even in dim corners in any temperature above freezing. Podocarpus, the somber Buddhist pine, forms dense pyramids of dark-green needlelike leaves; it also prefers cooler locations.

Among the many broad-leaved woody evergreens used as houseplants is Brassaia actinophylla, the umbrella tree, better known as Schefflera. Its spreading crowns of palmately divided, glossy green leaves do best in a light and warm location. Another picturesque plant is Polyscias fruticosa, the Ming aralia, with willowy, twisting stems densely clothed toward their tops with fernlike, lacy foliage.

The so-called rubber trees of the genus Ficus are widely used in homes and offices. All require good light to hold their foliage well. Best known is the large-leaved F. elastica 'Decora,' but perhaps even more attractive, because of their very graceful habit, are several small-leaved kinds, such as F. benjamina, F. retusa, and F. nitida. The giant violin-like, leathery leaves of F. lyrata, better known as F. pandurata, make the plant an attractive indoor "tree." Coccoloba, the sea grape, is another sturdy woody plant, somewhat resembling Ficus, with leathery, rounded leaves and crimson veining.

Because of their majestic beauty and distinctive decorative appeal many palms are grown indoors. Best known of the feather palms is theparadise palm (Howea, or Kentia), which combines grace with sturdiness; its thick, leathery leaves can stand much abuse. The parlour palms and bamboo palms of the genusChamaedorea have dainty fronds on slender stalks; they keep well even in fairly dark places. Similar in appearance is the areca palm (Chrysalidocarpus) with slender yellowish stems carrying feathery fronds in clusters. The pygmy date (Phoenix roebelenii), a compact palm with gracefully arching, dark-green leaves, is an excellent houseplant if kept warm and moist.

Ferns

Ferns, which come in a wide variety of forms, provide many popular houseplants. Among the best smaller parlour ferns is the sword fern,Nephrolepis, with bushy rosettes of leafy fronds; the holly fern (Cyrtomium), which has glossy dark leathery leaves; and the leatherleaf fern (Rumohra), with its leathery but lacy fronds. The bird's-nest fern (Asplenium nidus) forms a rosette of parchment-textured, fanlike, light-green fronds. Long-lasting Polypodium, often known as hare's-foot fern because of its pawlike, woolly rhizomes (rootlike structures), has feathery leaves on slender stalks. Among the attractive damp-loving ferns are the several species of dainty maidenhairs (Adiantum).

The so-called table ferns are a varied group of mainly Pteris and Pellaea species; some are frilly, others variegated; and in their younger stages they are ideal subjects for terraria. The Platycerium, or staghorn fern, has always aroused great curiosity because of its unusual shape. Growing as epiphytes on trees, these ferns have sterile fronds that cling snugly to the bark or, in cultivation, to a wire basket or wooden block; their much divided fertile fronds resemble the antlers of deer. One of the best of the palmlike tree ferns is the Hawaiian Cibotium, with a stout, fibrous trunk that bears a crown of light-green fronds.

Popular fernlike plants include Asparagus species that have plumy fronds. Species of Selaginella, called sweat plants or club moss, are strictly warm terrarium subjects; their delicate fronds greedily soak up moisture from the atmosphere to keep from shriveling.

Climbers and trailers

Climbers and trailers, weeping plants with stems too weak to support themselves, occur in most plant families. Best known are many varieties of ivy (Hedera). Generally, they prefer a cool location, but some small-leaved or variegated varieties do well on the windowsill. Several Cissus species, such as C. rhombifolia, the grape ivy, with metallic foliage, and the leathery C. antarctica, or kangaroo vine, are excellent plants for boxes or room dividers. Intriguing is the slow-growing Hoya, or wax plant, with leathery foliage and waxy, wheel-shaped blooms. By contrast, the inch plants and wandering jew, species of Tradescantia and Zebrina, are rapid growers with watery stems and varicoloured leaves; these long-beloved houseplants are used widely in window shelves or hanging baskets. The spider plants (Chlorophytum, or Anthericum) are houseplant favourites, forming clusters of fresh green ribbonlike leaves banded white; young plantlets develop from the tips of arching stalks.

Flowering plants

Most of the flowering potted plants seen at holiday times are not easy subjects for long-term indoor cultivation. They require high light intensity, careful watering, and day–night differences in temperature that are not usually available in the home; greenhouses offer better chances for successful cultivation. There are exceptions, however; one of the most successfully adapted houseplants is the African violet(Saintpaulia), with countless named varieties, with blossoms from violet blue through rose to white and single- and double-flowered forms. Window bloomers, such as Abutilon, the parlour maples, have bell-like flowers resembling Chinese lanterns. Impatiens, or busy Lizzie, is a genus of succulent herbs producing a succession of spurred flowers in gay colours. Hibiscus, the rose mallows, has short-lived giant blossoms in brilliant colours. Geraniums (botanically Pelargonium) have long been popular flowering plants in the sunny window; the foliage is often variegated or scented, and flower clusters may be in reds, pinks, and white.

A number of bulbous plants do well in lighted windows, including Hippeastrum, better known as amaryllis; Haemanthus, the blood lily; Neomarica, the apostle plant; and Veltheimia, the forest lily.

Orchids present a more difficult and specialized subject for successful home cultivation, usually because of their requirements for light, controlled temperature, and sufficient humidity and ventilation. There are some kinds, however, that give good results with ordinary care: epiphytic Epidendrumspecies, with waxy, usually fragrant, often greenish blossoms; and Oncidium species, or butterfly orchids, with brightly coloured, long-lasting yellow flowers marked with brown and often produced in large sprays.

Small flowering plants that produce edible fruit can be grown on a windowsill. With sufficient light and ventilation, success may be had with the Calamondin orange (×Citrofortunella mitis), the dwarf Chinese lemon (Citrus limon 'Meyeri'), and the American-wonder lemon (C. limon'Ponderosa'). The fig tree (Ficus carica) can be grown to yield edible fruit, as can the dwarf Cavendish banana (Musa acuminata, formerly M. nana) and the dwarf pomegranate (Punica granatum nana), the pineapple (Ananas comosus), and the coffee tree (Coffea arabica).

FERTILIZING HOUSEPLANTS

Through the process of photosynthesis, plants manufacture the sugars and carbohydrates needed for their growth and development. They require at least 16 elements to carry on this process. Plants obtain carbon, oxygen and hydrogen from air and water. Outdoor plants would obtain the other nutrients required for growth from the soil. Houseplants rely on us to supply these nutrients through the application of fertilizers. Underfertilized houseplants often exhibit symptoms of slow growth, weak stems, pale leaves and reduced flowering.

TYPES OF FERTILIZERS

Houseplant fertilizers come in a number of formulations. Some are wettable powders or concentrated liquids that are diluted with water then applied to the potted plant. Time release fertilizers are available as coated pellets or as spikes. A few are sold premixed and applied directly to the potting soil.

The label on a fertilizer container states the guaranteed analysis or grade. These are the three numbers listed on the package (i.e. 8-7-6). They refer to the percentage of nitrogen, available phosphorus and water soluble potassium contained in the fertilizer. Plants require large amounts of these three nutrients and they are often referred to as primary nutrients or macronutrients. Nitrogen promotes green, leafy growth. Phosphorus encourages flowering and root growth, and potassium is necessary for stem strength and stress tolerance. Secondary nutrients that are required in slightly lesser quantities include calcium, magnesium and sulfur. Ground limestone is sometimes added to potting mixes to supply calcium and magnesium.

The fertilizer label will also state from what sources these nutrients are derived; chemical and/or organic. If synthetic or chemical fertilizers contain trace elements, these too will be listed on the label. Trace elements or micronutrients are needed by plants in very small amounts and include iron, boron, zinc, copper, molybdenum, chlorine and manganese. It is usually safe to assume that organically derived fertilizers, because of their nature, contain some or all of the necessary micronutrients. Potting mixes that have a mineral soil component may also supply sufficient amounts of micronutrients. If you are not using an organic houseplant fertilizer, consider purchasing a chemical fertilizer that does include micronutrients since they are typically not present in soilless potting mixes.

Choosing a Fertilizer

Which is the best houseplant fertilizer for you to use? This will depend on the types of plants being grown, cultural conditions and your schedule. In general, foliage houseplants appreciate fertilizers high in nitrogen while flowering plants respond best to those with higher phosphorus analysis. There are plenty of specialty houseplant fertilizers out there but do examine their labels. Often the difference is more in the packaging than in the amounts or proportions of nutrients supplied.

Purchase a water-soluble powder or liquid concentrate if plants are to be fertilized on a weekly, biweekly or monthly basis. If there will be long intervals between fertilizer applications, select time release formulations in either pelleted or spike forms. These can be applied at intervals from 2 to 9 months and will provide houseplants with a steady supply of nutrients.

Frequently houseplant lovers amass quite a collection of different plant species. Sometimes plants have specific fertility requirements but usually an all-purpose, balanced fertilizer could be applied to all plants. For example, a fertilizer where the three numbers on the package are equivalent or just about equal, such as a 20-20-20 or a 10-8-7.

When to Fertilize

Houseplants respond to fertilizer during periods of active growth. This is usually from March until October. Reduced light and temperatures throughout the winter months often render a plant inactive and it is generally recommended that plants not be fertilized at this time.

The labels on most water-soluble fertilizers recommend monthly applications. Since these nutrients are easily leached from the potting mix, your plants may benefit from more frequent dilute pplications. If one teaspoon per quart of water is recommended for monthly feedings, you could feed bimonthly using only one-half a teaspoon per quart or weekly using a quarter teaspoon per quart. This gives the plant a steady, continuous supply of nutrients. This type of regime is often recommended for flowering plants like African violets.

When fertilizing houseplants always follow the directions on the label. More is not better and excess nutrients can harm roots and leaves. Always apply fertilizer to an already moist potting soil to avoid root damage.

Overfertilization

Browning roots and leaf tips, wilting, poorly shaped leaves and a white crust on pot rims may indicate overfertilization. Excess nutrients in the potting soil will desiccate or burn tender roots. High concentrations of nutrient salts also prevent the plant from taking up water so wilting is observed.

Often leaching the potted plant with copious amounts of water will reduce excessive fertilizer salt levels. Be sure water can drain freely. Another solution would be to repot the plant, gently removing as much of the old potting mixture as possible, and replacing it with fresh medium.

Keep houseplants healthy and thriving by practicing good watering practices, meeting their light and temperature requirements and providing adequate nutrition through a regular fertilization program.

REPOTTING HOUSEPLANTS

As your houseplant grows larger and the roots either begin to grow through the drainage holes or become pot bound, repotting the plant into a larger pot will become necessary. After deciding to repot, following a few steps is all that's needed to complete this task successfully. First though, have fun looking for and choosing a container that will really complement your houseplants

As your houseplant grows larger and the roots either begin to grow through the drainage holes or become pot bound, repotting the plant into a larger pot will become necessary. After deciding to repot, following a few steps is all that's needed to complete this task successfully. First though, have fun looking for and choosing a container that will really complement your houseplants.

CHOOSING A CONTAINER

Many varieties of containers are available, but the container you eventually select should be based on your preference. There are, however, a few considerations to think about when choosing a container. The new container should be just slightly larger than the current one, as a container that is too large may look out of balance with the overall plant. Containers should not detract from the plant; rather the size, color, and material chosen should complement the size of the plant, leaf texture, and leaf and flower color. Before repotting, test perspective containers by placing potted plants inside them. Then place the plant in the location where you'll display it permanently, step back, and take a look at the entire combination. Ask yourself whether the color and texture of the container compliments or detracts from the plant. Is the size of the container appropriate, or is it too small or too big for the plant? If you are not pleased with the look, try another option and give the combination another critical look.

When choosing a container, the basic options are plastic and clay (terra-cotta), clay being available in glazed and unglazed. Both have advantages, and again the option you choose is primarily based on your preference. An advantage of choosing a plastic container is that it is lighter weight and easier to move than a clay pot when the plant and potting mix are added. Clay pots, on the other hand, are less likely to tip over when a plant becomes top heavy and the potting mix dries out.

Other advantages of using clay pots include reduced waterlogging and salt buildup, as the clay tends to pull water and soluble salts from the potting mix. For these same reasons, plants in clay containers will need to be watered more often than plants in plastic containers. The decorative features of these types of containers can be numerous, as more ceramic, fiberglass, and other synthetic containers are available for purchase.

Another option is to place potted houseplants in a potholder. Potholders can be made of metal, wood, or other decorative materials; however, they may not be suitable to plant a houseplant in. Instead, a potted houseplant can be slipped inside the container to achieve the desired look.

A container with drainage holes is necessary for excess water to drain, preventing the potting mix from becoming waterlogged—this can contribute to potential disease problems. Drill holes in the bottom of a container if necessary. Then use a drip tray to collect excess water and prevent furniture or surfaces that are not waterproof from becoming damaged. If it is not possible to drill a hole through a decorative container, first repot the plant into a container with a drainage hole and then use the decorative container as a potholder. Make sure that there is space between the drainage hole and the bottom of the decorative container to allow water to drain freely from the potting mix.

Deciding if the plant is pot-bound

The best time to repot a plant is in the spring so that actively growing roots will have enough time to grow into newly added potting mix. There are several signs that houseplants can exhibit when they are pot-bound. First check the frequency you are watering the houseplant.

- Do you notice that the potting mix dries out quickly even during periods of cooler temperatures?
- Also, are any roots growing through the drainage holes?
- Does it appear that the foliage and stems have either stopped growing or growth has slowed even when fertilized regularly in spring and summer?

If the answer to any of these questions is yes, remove the plant from the container and examine the roots. If the roots are circling around the rootball or are so dense that it is difficult to see any potting mix in the bottom third of the root ball, it's time to repot.

When you discover that your houseplant is pot-bound, you have a few options. First you need to decide whether you want the plant to continue to grow larger or if you're satisfied with the current size. If you don't want a larger plant, it will be necessary to remove the outer section of plant roots, return the rootball to the container with some new potting mix, and cut back some of the plant's top growth. If the goal is to let the plant grow larger, get ready to repot it.

Choosing potting mix

Using garden soil in indoor containers for houseplants could lead to disease and pest problems. Instead, choose a potting mix or potting soil especially created for potted plants. These could include a composted soil or peat mixture with fertilizers to supply nutrients. A peat-based potting mix will weigh less than a soil-based mix; however, it is more difficult to wet if it dries out. Wet the potting mix prior to repotting houseplants to ensure that the potting mix will absorb water evenly.

Repotting the plant

After you chose an appropriate container and potting mix, you can begin the repotting process. Water the plant in its original container and let it sit for one hour before repotting. If the container was used previously make sure that it is clean before you start. If you have a new plastic container, no preparation is necessary. A new clay container will require overnight soaking in water. This will prevent the container from soaking up any moisture from the potting mix when the plant is first repotted. If using a clay container with drainage holes, place a few pieces of a broken clay container over the hole. This will allow excess water to drain through the hole, but prevents potting mix from doing so. Next, remove the plant and gently "tease" the roots so that they are no longer circling the rootball or are densely matted. Place some potting mix in the container so that when the rootball is placed on top, the top of the rootball is slightly below the lip of the new pot.

Then add more potting mix around the rootball and gently firm the top layer so that the potting mix fills the container, but is not so compacted that it restricts water and air movement into and through the pot. When filling the container, leave an inch of space between the potting mix and the top of the container so that there is room for water and additional plant growth. Next, water the potting mix and place it its intended location.

3

Designing of Indoor Gardening

WHY GARDEN INDOORS?

In an outdoor environment, nature gives the plant what it needs to grow: sunshine to power photosynthesis, mineral elements to convert into starches and sugars, water to carry food through the plant, and carbon dioxide for plant respiration. But some of these provisions may be more or less than the plant needs, so the resulting environment is seldom "ideal." In a controlled indoor growing environment (whether a large greenhouse or a small home grow room), you - the grower - must sup-ply all of what the plant needs. In the beginning, this method may seem like more work than gardening outdoors, but the ability to control all of the plant's growth factors can result in much greater yields in a smaller amount of space which ultimately means LESS work for you!

INDOOR LIGHTING

Lighting is the most important factor in indoor gardening, and often the biggest financial investment. Outdoors, the sun provides more than enough light energy for plants, but growing seasons and conditions vary widely in different parts of the country. Plus, many people simply don't have the space for an outdoor garden. As long as you have adequate light, an indoor garden can provide you with fruits, vegetables and herbs year round...regardless of what is happening outside.

The main types of lighting for indoor gardens are fluorescent and High Intensity Discharge (HID).

Fluorescent Lights

Fluorescent lights are excellent for starting seeds or rooting clones. Fixtures and bulbs are inexpensive, and the low heat output lets you put them just inches away from your delicate plants. The disadvantage of fluorescent bulbs is the low light intensity—they can grow a plant that is 8-10" tall but then the light simply can't penetrate any further. If the plant grows taller and you keep raising the bulb, lower sections of the plant will not receive adequate light. Using a "full spectrum" fluorescent bulb will give your plants all the necessary wavelengths of light.

High Intensity Discharge (HID) Lights

High Intensity Discharge, or HID, light bulbs have revolutionized the indoor gardening industry in the last 25 years. These bulbs require spe-cial ballasts (transformers) and sockets to operate, which are included with light systems. HID light systems are designed to operate on normal household (110v/120v) current, but most can be converted to 220/240 volt operation if desired. These lights are designed for use in home gar-den rooms and greenhouses, and are completely safe if used with com-mon sense and according to the instructions.

Many types of light bulbs fall into the HID category, but the two best choices for plant growth and maintenance are Metal Halide and High Pressure Sodium.

Metal Halide (MH/MS)

Metal Halide lamps are the best single source of artificial light for indoor gardening. Their balanced light spectrum, similar to the tropical sun, contains the important blue and red wavelengths that plants need for rapid vegetative growth. You can grow your plants from start to finish under a metal halide lighting system. MH lamps come in 175, 250, 400, and 1000 watt sizes. NOTE: Horizontal Metal Halide light systems require the use of a tempered glass safety lens in order to be UL listed for safety.

High Pressure Sodium (HPS)

High Pressure Sodium lamps emit light which is heavily concentrated in the red and orange region of the spectrum. This heavy red light pro-motes excellent fruit or flower production (as much as 30% more than a metal halide lamp), but the lack of blue spectrum light can sometimes make a plant stretch or become "leggy" during the vegetative growth stage. This type of light is ideal for supplementing sunlight in a green-house or sunroom. A High Pressure Sodium system is the most efficient HID light because it has the highest number of lumens per watt (roughly 10-15% more than a metal halide bulb of the same wattage.) HPS lamps come in 150, 250, 400, 600, and 1000 watt sizes. The ideal light-ing system would include both a Metal Halide and a High Pressure Sodium lamp. This would produce extremely fast growth as well as increase flowering by 40 percent. To avoid the expense of purchasing two separate lighting systems, you can use a conversion bulb or an enhanced spectrum bulb.

Conversion Bulbs High Pressure Sodium Conversion bulbs are specially designed to run off a Metal Halide ballast but they put out more lumens, more red spectrum light, and they run off less electricity. Start your crop under the MH bulb and then switch to the HPS conversion bulb when flowering begins.

Bulbs that convert MH into HPS are available for 175, 250, 400, and 1000 watt systems. Metal Halide Conversion bulbs allow you to take the opposite approach: grow your plants under the conversion bulb during the vegetative stage, then switch to your regular high pressure sodium bulb for flowering. HPS to MH conversions are only offered in the 250, 400, and 1000 watt sizes.

If you are planning to purchase a system with a conversion bulb, con-sider which light spectrum you will use most. If you will need mostly the vegetative light spectrum with short periods of flowering light, buy the metal halide system with a high pressure sodium conversion bulb. If you will have a relatively short vegetative phase followed by a longer flower-ing phase, buy the high pressure sodium system with the metal halide conversion bulb. Either approach will work, but purchasing the system of the type you will use most will give you much more efficiency in the long run.

Enhanced Spectrum Bulbs

If you only want to deal with one bulb through all stages of growth, another option is to use an enhanced spectrum or "corrected" bulb. High Pressure Sodium Son Agro bulbs are engineered to provide 30% more blue spectrum light than a standard HPS bulb. These are available in 160, 270, or 430 watts. Agrosun Metal Halide bulbs are engineered to provide more red spectrum light (38%—49%) than a stan-dard halide bulb. Agrosun bulbs are available for 250, 400, and 1000 watt systems.

Bulb Replacement

Replace metal halide bulbs after 1 to 1 1/2 years of use, and replace high pressure sodium bulbs after 1 1/2 to 2 years of use. The bulbs will continue to light beyond this point but have lost as much as 30% of their lumen output while still consuming the same amount of electricity. When replacing your bulb, it is critical to get the right one for your ballast and reflector configuration. Contact AHL if you have any questions!

Light Movers

You can use an automatic light mover to increase the coverage area and efficiency of your light without using any more electricity. Since the light is always in motion and does not rest above any part of the gar-growing season.

HYDROPONIC GROWTH MEDIA

By definition, a hydroponic medium should offer no nutritive value-its sole purpose is to provide support for the plant and to allow an even distribution of the nutrient solution. An ideal growing medium is sterile, inert, water- and nutrient-retentive, free-draining, non-toxic, and will pro-mote vigorous root development.

Rockwool

Rockwool is the most revolutionary soilless medium to date. Horticultural Rockwool is made of volcanic basaltic rock and a binder which prevents potassium hydroxide and other elements from leaching into the nutrient solution. Rockwool provides 90-95 percent air space between its fibers. It is capable of holding more nutrient solution and more air than any other medium. When the Rockwool is completely sat-urated, it maintains a ratio of air to water which is ideal for promoting root development: 80 percent nutrient solution, 15 percent air space, and 5 percent Rockwool fibers.

Fired Clay Grow Rocks

Fired Clay Pellets, often called grow rocks, are manufactured by heating clay pellets to very high temperatures, causing them to expand and puff up with air. Grow rocks retain moisture because they are porous, while the irregular shape and rough surface of the rocks promotes free drainage and air circulation. Grow rocks can also be reused from one crop to the next.

Coconut Fiber

Coconut Fiber is an exciting new hydroponic medium that also works great as a soil amendment. The fibers, which are also called coconut coir, are made from the inner husk of coconuts. This renewable resource offers excellent water holding capacity along with free drainage. Coconut fiber is rich in natural rooting hormones and other organic compounds.

Perlite

Perlite, or puffed sand, is a sterile and lightweight medium. It is a good medium for lettuce, cabbage, herbs, and other small crops. Its major drawback is its inability to give good support to the plants. It also tends to promote algae growth more than other media. Perlite is an excellent medium when used in grow bags or when combined with clay pellets in a reservoir or wick system. Perlite should be discarded after only one use.

PLANT FOODS & NUTRITION

All plants need nutrients. Outdoors, plants take certain elements from the earth and combine them with water and air. Using energy from sun-light they create simple sugars, carbohydrates, and proteins to feed both themselves and all living things on this planet. A hydroponic solu-tion must supply all the necessary nutrients in the proper form and ratio for plant growth. All major, minor, and trace elements must be properly balanced and water soluble. The three major plant elements are nitro-gen, phosphorus, and potassium—often abbreviated as N-P-K. The three numbers on every fertilizer label refer to the percentages of these three essential elements. The minor elements include calcium, magnesium, sulfur, and iron. The trace elements necessary for good plant develop-ment are manganese, boron, copper, zinc, chlorine, molybdenum, and cobalt. Plants need very small amounts of the trace elements, but they are important because they act as catalysts in the solution. Without the right amounts, a plant can't properly utilize the major elements, and may develop deficiencies, disease, reduced growth rates, and poor yield.

Synthetic Nutrients

It is important to use a nutrient that is designed specifically for hydroponic applications. While products like Peter's and Miracle Grow are adequate for soil plants, they will not perform well in hydroponic sys-tems because they lack the proper ratios of trace elements. If you are using a dry or solid nutrient concentrate, make sure it's derived from soluble and chelated minerals. Most hydroponic gardeners use a syn-thetic (or chemical) liquid nutrient formula, and many excellent brands are available. These liquid nutrients contain all major, minor, and trace elements in exact proportions, and they are usually formulated to keep pH stabile and reduce fertilizer salt build-up. Synthetic nutrients will pro-vide the fast, lush growth that most people associate with hydroponics!

Organic Nutrients

Think of organic nutrients as a raw food source for plants. While a synthetic, or chemically derived, nutrient will contain mineral elements (such as nitrogen, calcium, etc.) that are immediately available to the plant, an organic nutrient (such as bat guano or bone meal) must break down in order to release these mineral foods to the plant. Organic nutri-ents are excellent for use in soil because beneficial bacteria will speed up this process. In hydroponics, organic nutrients become trickier because these living "catalysts" are not usually present. However, many organic nutrients can be used successfully in hydroponic systems, and some gardeners find that the slower growth rates are worth it for better looking, better tasting produce. Depending on the type of hydro system you use, you will have to decide on the best type of organic nutrient. Consider if clogging will occur (drip emitters, lines, pumps, etc.), and also the different smells associated with organic fertilizers. An organic/synthetic combination will promote the aggressive growth asso-ciated with synthetics and the high quality and heavy yield standards of organically grown produce. Using organic nutrients or supplements will improve the overall health, taste, and texture of all types of plants.

Nutrient Cycles

A plant's nutritional requirements change depending upon its stage of growth. Generally, during the vegetative cycle a plant requires high amounts of nitrogen with lesser amounts of potassium and phosphorus. During this stage you should use a Grow formula nutrient. When the plant moves into the blooming or flowering stage, the plant needs more phosphorus and much less nitrogen. During this stage, use a Bloom for-mula nutrient instead of the Grow formula in order to slow vegetative growth and promote flowering and fruiting. You should also adjust nutri-ent strength depending on the age of your plants. Feed seedlings and cuttings with a mild strength nutrient solution for the first 2-4 weeks, then increase to a "normal" strength nutrient mixture. After the first month of growth is completed, you can determine the nutrient solution strength according to the needs of your plants. It s best to follow direc-tions on the label, or contact AHL with any questions. It is very easy to over fertilize, and difficult for plants to recover from fertilizer burn. Remember, when it comes to nutrients, less is more!

pH

The pH reading is a measure of a solution's acidity or alkalinity. It is measured on a scale from 1 to 14: an acidic solution has a pH less than 7.0, while an alkaline solution has a pH greater than 7.0. The pH of a nutrient solution or soil is directly related to the plants' ability to absorb the necessary nutrients. Most plants perform best in a hydroponic solu-tion maintained at an average pH of 6.3. Your pH will fluctuate constant-ly, so check and adjust it daily if possible. You can adjust pH by adding either pH up (base) or pH down (acid) to the water. Household chemi-cals, such as vinegar, are not recommended because they cannot main-tain a stable pH after adjustment. To adjust pH of water already circulat-ing in your system, add the acid (pH down) or base (pH up) to your sys-tem, then check pH again in about one hour—after the adjustor has fully mixed and stabilized. For a soil garden, it is not as critical to check and maintain an exact pH because soil is a better buffer for nutrients than a hydroponic solution. However, many soil gardeners will check pH peri-odically just to make sure everything is OK, and soil gardeners should check pH if the plant shows signs of nutrient deficiency that might be related to pH imbalance.

Total Dissolved Solids (TDS)

A dissolved solids reading, measured in terms of parts per million (PPM), tells you how many parts of a dissolved solid are suspended in a solution. (A reading of 1500 ppm means there are 1500 solid parts dissolved in 1,000,000 parts of solution.) You should maintain your nutri-ent solution between 600 and 1200 ppm, although ideal levels change from plant to plant, and even from one growth stage to the next. If your dissolved solids reading is too high, add plain water to the solution to dilute the salts. If your TDS reading is too low, add nutrients to increase the concentration in your solution. A TDS reading is a helpful way to judge overall nutrient strength, but it cannot distinguish between differ-ent elements, or between dissolved solids that were present in the solu-tion before you added the nutrient. Dissolved solids can only be measured using an electronic TDS meter.

Electrical Conductivity (EC)

Electrical Conductivity (EC) is another way to measure the relative strength of a nutrient solution. (In fact, a TDS meter is actually measur-ing the electrical conductivity of a solution but converting that data inter-nally before displaying a reading in PPM). As with TDS, the only way to measure EC is using an electronic EC meter. One advantage of using an EC meter is that EC is the standard of measurement in every part of the world other than North America. Another, and more important advantage, is that EC readings are more consistent than TDS readings because different companies use different conversion factors to achieve the TDS reading. Therefore TDS meters from different companies will not necessarily give the same readings in the same solution.

Water Temperature

You should maintain a constant temperature between 70¡ and 80¡F in your nutrient reservoir. This is important, especially during the cool months, to help increase plant performance. Do not increase the tem-perature above 85¡F as this may cause root damage. You can use an aquarium heater to maintain the temperature in your reservoir. It takes at least 5 watts per gallon to heat and maintain a constant nutrient tem-perature (for example, a 10 gallon reservoir requires a 50 watt heater). You can also place the reservoir on a piece of Styrofoam or wood to provide some insulation if you are growing on a concrete or tiled floor.

ENVIRONMENT

Air Temperature & Humidity

Most plants grow well with room temperatures between 50¡ and 90¡F, with a median temperature of about 75¡F. Your specific plant's needs will vary: think about where your species of plant originates and try to recreate a similar environment. Plants also need a certain amount of water vapor, or humidity, in the air to help control transpiration (breath-ing) and prevent wilting. On average, a relative humidity between 25 and 75 percent is good, with the median range around 50 percent.

If your room is too hot or too humid for you, it is probably too hot and humid for your plants. Excess heat and humidity should be vented away from the growing area with exhaust fans. Alternatively, you can add a heater if the air is too cold, or a humidifier if the air is too dry. Environmental control equipment can be used to turn on fans when temperature or humidity rises, then turn them off when levels have dropped. You should also use a small circulation fan in your garden area. This fan can run all the time to provide air movement around the plants, strengthening plant stems and providing fresh air to the stomata (the cells on the undersides of leaves where the plant "breathes").

Carbon Dioxide (CO_2)

Carbon dioxide (CO_2) is probably one of the most overlooked require-ments for good plant growth. Plant respiration is opposite of human res-piration- they breathe in carbon dioxide and breathe out oxygen. Normal air contains 300-400 ppm of CO_2, but growing plants quickly use this amount when confined to a small space.

Test kits are available to deter-mine how much CO_2 is in your air. If you exchange your air regularly—by using a vent fan to bring in fresh air and another fan to exhaust stale air—you can usually give your indoor plants a CO_2 level similar to what they would have outdoors. However, if you add much greater amounts of CO_2 to the air you can speed up photosynthesis, resulting in faster growth and greater yields. For every increase in CO_2 up to 1500 ppm, there is an increase in growth rate.

This is called the "point of diminishing return," after which point each increase causes a corresponding decrease in growth rate. You can add CO_2 to your grow room by installing a CO_2 tank with a regulator and a solenoid valve, then attach this unit to a timer to disperse measured amounts of CO_2 at regular intervals.

Always place distribution tubing above your plants because CO_2 is heavier than air and "falls" as soon as it leaves the tub-ing. Another option is to run the CO_2 tubing into the back of a small oscillating fan which will blow CO_2 through the plant canopy. You can also introduce CO_2 using special generators which run off natural gas or propane. These generators produce quite a bit of heat, and they must be located directly inside the garden area, so keep this in mind when planning the garden environment.

STARTING PLANTS

Starting Seeds

"Plant propagation" simply means starting plants from seed or clones. Most indoor gardeners prefer to start their own seeds rather than buy seedlings from a nursery. They are usually able to get stronger, healthi-er plants because they can optimize conditions from the very beginning, plus eliminate the possibility of bringing in an insect or disease. Plus, you can buy exotic seed varieties from all over the country, or even other parts of the world. Gardening is a lot more fun when you aren't limited to the two kinds of boring tomato plants on sale at your local nursery! Hydroponic gardeners benefit the most from starting their own seeds because they can sow them in a soilless medium such as rock-wool, which makes the transition to a hydroponic system much easier.

Cloning

The word "cloning" may sound complex and scientific, but the idea is as simple as taking a cutting from a houseplant and rooting it in order to produce another plant. The advantages of cloning are numerous: you can select your best specimen to be the mother plant, meaning that your clones will have the same good genes and growth characteristics; you can speed up your time to harvest because clones are ready for transplanting much faster than plants grown from seed; and clones tend to be more uniform in size and height than seedlings—a big advantage if you are gardening with artificial light.

INSECTS & PESTS

Indoor gardens tend to have fewer insect problems than outdoor gar-dens, especially if hydroponic systems are used. Many insects spend part of their life cycle in soil, so eliminating that also eliminates those insects. In addition, hydroponically grown plants tend to be stronger and healthier than soil-grown plants, which makes them more resistant to insects. But of course problems do occur sometimes, especially since tiny pests can enter your garden room from other parts of your home, through vent fans, or even on your clothes or shoes! Pests are usually controlled more easily inside than outside because the space is con-fined, but they may also multiply rapidly without the natural predators that keep their numbers under control outdoors. If you find insects in your indoor garden, there are many ways to eradicate them or at least keep them under control. Most home gardeners prefer milder pesti-cides, such as those made from the pyrethrum plant, or "insecticidal soap" sprays which kill insects on contact but leave no residue on plants. Plant oil extracts, including neem oil, offer a totally organic method. And more and more growers are using natural predators and parasites. These "beneficial insects" are raised in specialized insec-taries and are the most "natural" form of pest control available.

PLANT DISEASES

Plant diseases can be placed into one of three categories: fungus, bac-terium, or virus. Most plant diseases are caused by fungi. Many plant diseases are caused or encouraged by poor drainage, poor or unbal-anced soil/pH/ nutrients, inadequate air circulation, insect damage, or unsanitary conditions. When plants show unhealthy signs, analyze the symptoms. Before assuming it is a disease, first look carefully on the undersides of leaves for a pest problem. Typical symptoms of disease include spots of various sizes and colors, abnormal localized swelling (galls), blights (sudden death of foliage, branches, or flowers), rots (gen-eral decomposition of plant tissue), cankers (dead areas on bark or stems which are often discolored and may be raised or sunken), and general dwindling of plant health. There are a few things you can do to help prevent plant diseases. Whenever possible, work in your garden area when foliage is dry. Most bacteria and fungus need moisture to travel from plant to plant. Make sure to wash your hands after removing any diseased plant from the garden. You can also plant disease-resist-ant plant varieties. Make sure to allow enough space between plants for the air to circulate freely. Fungus can be prevented by controlling tem-perature and humidity. A fungus can spread like wildfire! Use a fungicide spray if the fungus gets a good start and appears to be spreading. Apply the fungicide at least twice, about 5-10 days apart.

PREPARATION AND DESIGNING OF INDOOR GARDENING

Decide on a style of garden. When gardening indoors, there are two general styles of gardening: container gardening, and hydroponic (also called aquaponic) gardening. Container gardening is just as it sounds - a series of planters/ containers that use traditional gardening soil and methods to grow your plants. A hydroponic garden is a special type of indoor garden that uses fertilized water (sometimes with live fish) and a non-soil starter for your plants, arranged vertically. Each is a valuable option for different reasons:

- Container gardens are a good option for someone who may want to rearrange their plants or eventually move them outdoors. Container gardens can grow any type of plant, in any size.

- Hydroponic gardens are a good option for someone who is looking to produce a lot of plants in a small amount of space. Typically, hydroponic gardens are used for growing vegetables.

Choose the right space. Creating a successful indoor garden relies heavily on selecting a space indoors that will help your plants to grow. Choose an area that has lots of windows and sunlight; typically east and west facing windows are the best. When at all possible, place your garden (container or hydroponic) near a window to allow for more heat and sunlight.

- Avoid rooms that have cold temperatures (such as an attic or garage); cold can kill or slow the growth of your plants, while heat is typically more universally welcomed by your plants.

- Avoid choosing an area near an air vent or fan, as these can dry out your plants and cause damage to them.

Properly control the environment. The biggest reason for planting an indoor garden is the ability to completely control the environment. While this requires more work, it also greatly increases the success of your plants when done properly. There are three general things you'll be in charge of controlling: air temperature, water frequency, and soil condition. These will vary slightly depending on the system of garden that you are using and the plants you decide to grow, but there are a few helpful tools to make controlling the environment easier.

- Try soil heat mats. In general, most plants thrive in soil temperatures between 75–85 °F (24–29 °C). You probably don't want to have your house heat on that high, especially in the winter. To remedy this, you can purchase electric mats made for heating pots from the bottom, regulating the temperature of the soil.

- Get a drip system. Watering regularly can be a difficult habit to master; rather than setting reminders for yourself daily, try getting a drip system. This works with a series of small tubes supplied to each plant, and a timer that turns the water on/off for a certain period each day.

- Choose a lighting system. Although windows let in a little sunlight, you'll primarily be in charge of giving your plants light through artificial means. Fluorescent light tubs are typically considered the best choice, because they are cost effective and give good results. Otherwise, you can buy a specialty heat-lamp for growing indoor plants.

Choose your plants. Just like a garden outside, there are a huge variety of plants you can grow indoors including vegetables, berries, herbs, flowers, and decorative greens.

Before you head to the nursery and start picking out random attractive plants, you must determine which plants will work well in your garden. Some plants will flourish indoors, while others will struggle. Some popular plants to grow indoors include:

- Vegetables: tomatoes, peppers, eggplant, carrots, radishes, beans, peas, and mushrooms.

- Fruits: strawberries, figs, and grapes.

- Herbs: basil, bay, chives, oregano, parsley, rosemary, sage, tarragon, and thyme.

- Flowers: peace lily, african violet, marigold, begonia, cactus, and succulents.

Choose your containers. Because you're growing indoors, you have a wide variety of options for containers to plant in. You can go with traditional pots or containers from a garden supply store, or be unique and use old vases or plastic bottles.

You don't need a large container to start seeds, and if you're transplanting a plant you'll need a container twice the size of the root ball. Otherwise, look for a container that has holes in the bottom to allow for drainage, or simply drill holes into any container.

- Plastic containers retain moisture the best, but terra cotta pots are typically seen as the most attractive option.

- Try recycling old coffee cans or 1-liter plastic bottles to use. This way you'll be green while gardening - a double bonus!

- You can line the bottom of your container with rocks for added drainage.

- If you use a wooden container, try to find one made of redwood or cedar, which are particularly rot resistant.

- Don't use any container that has been treated with chemicals, as this could kill your plant.

Create your potting mix. Unfortunately, you can't just grab a bunch of soil from your garden and use it for planting your containers. Soil from outside often contains diseases and insects that can kill off your plant over time, and is rarely the proper balance of sand/clay for the ideal porous-ness. While you can purchase pre-made potting mix, you'll save more money and guarantee the quality of your mix by creating your own. To make your own potting mix, you'll need 1 part coir peat, 1 part vermiculite, and 2 parts compost. These are all available at your local gardening center.

- Soak the coir peat brick to rehydrate it; typically these come with directions, so follow the attached instructions on soaking.

- Mix your coir peat and vermiculite together until well mixed, and then add in the compost.

- If you're able, worm castings are an excellent addition to your soil mix; add ½-1 cup of worm castings to your mix prior to planting your containers.

Set up your garden system. If you're planting a large indoor garden, you'll need to create a shelving system for your plants. Otherwise, use the aforementioned tips to select a space for your container garden. You'll also have to set up your control factors: your light, water system, and temperature controls. If you're using a shelving system, it should be easy to hang fluorescent lights and your drip system to the shelves. Otherwise, work your space until you have all the components adjusted. Heat mats should be placed underneath the locations of your containers.

- You can purchase timers for your lights, heat mats, and drip systems so that they are only on during certain times of the day.

- Keep in mind that certain plants require different amounts of light, and group them appropriately. For example, put all the light-loving plants near each other and the shade-loving plants near each other, and adjust the light output accordingly.

Maintain your plants. After planting your garden, you'll need to upkeep it in order to be successful. Make sure that your plants are getting regular amounts of sunlight and water, and that the soil temperature does not drop below 70 °F (21 °C). When your plants become too large, you'll need to transplant them into larger pots or split them into more plants, either to keep or to give away.

- If you notice any plants with brown spots, that are wilting, or are clearly dying, remove them from the rest in case they carry a disease or pest that might spread.

- You can incorporate compost or fertilizer into the containers every few months to help supply the plants with nutrients.

- You can set your pots in trays of rocks, so that the water that drains out will provide humidity for them as they grow.

Part 3 of 3: Hydroponic Gardens

Choose a hydroponic system. There are many different ways to set up a hydroponic garden, but they all follow the same theme: a water tank is located on the bottom, and stacks of shelves/plants are placed on top. The fertilized water from the tank is fed to the plants using a pump system, and the water filters down through the plants and excess is drained back into the original water tank.

A few variations of this system include:

- Using fish in your tank. Freshwater fish will naturally fertilize the water through their waste, creating an ideal mixture for your plants.

- Creating a small-scale hydroponics system. You don't need a massive shelving unit and lots of supplies to create a basic system. Instead, you can use cut-off plastic bottles and a sunlit window. Consider making a window system for a tight space.

- Using one large container instead of many small ones. There are so many variations of hydroponic systems, and one of them includes using a large tub to produce a lot of the same crop, instead of using multiple small containers.

Choose your location. Just like with a container garden, a hydroponic garden will thrive with the most available sunlight. Choose a space near a window; in fact, if you're using a small garden the window may be the only necessary source of light. Otherwise, you'll need to find a location with a lot of vertical space for the stacking that is necessary for the system. Try using a shelving unit or organizational bookshelf for your hydroponics system.

- Don't place the shelves directly over or under an air vent/duct.
- When possible, put your hydroponics system on hard floors rather than carpet to avoid mold and mildew growing.

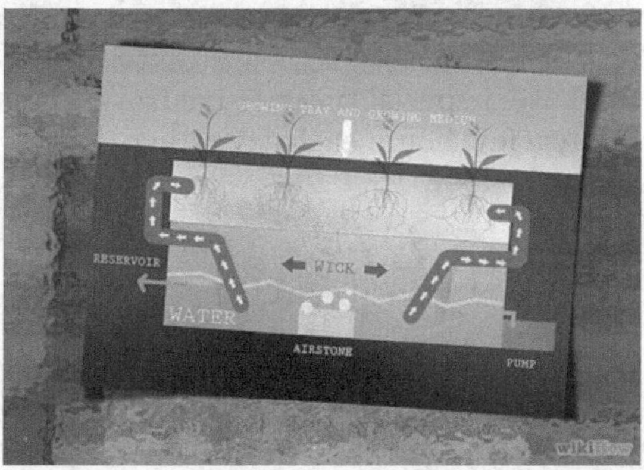

Prepare your containers. As with a container garden, virtually any container is appropriate for a hydroponics system so long as it isn't chemically treated and allows for drainage through the bottom. You'll need to drill holes into a container that does not have them, so that the water can filter out at the bottom. Instead of using potting soil for your containers, though, you'll be filling your pots completely with perlite. Perlite is a type of porous soil replacement that works well for this type of compact garden. Then, plant your seeds as per usual and water them heavily to help reduce transplant shock.

4. Set up your system. If you purchased a commercial hydroponics system, you'll need to follow the package directions for setting it up. If you're making

your own homemade system, you can follow these loose directions for set up. Put your water tank on the bottom shelf, elevated at least a few inches above the floor. Then, place your plant containers on the shelves above the tank; the container closest to the tank should be nearly covering it completely - there shouldn't be much distance from the top of the tank to the bottom of the container. Set up your water pump system to deliver water to the plants on the shelves above.

- Your fluorescent lights should be hung directly above the plants.

- You can purchase a special timer to control the water pump, so that it goes for only a few hours each day rather than 24/7.

- Talk to a local hydroponics expert at a nursery for specific advice to your system.

Maintain your system. Over time, you'll need to make sure that your hydroponics system is in good condition so that it keeps running smoothly. Replenish water in the tank if necessary, and adjust the position of the lights as the plants grow. If you used fish in your tank, you'll need to make sure they are in good health and aren't lacking food or space.

ESSENTIAL EQUIPMENT FOR INDOOR GARDENING

GROW LIGHTS

For indoor gardening, lighting is paramount. Without adequate light, plants become leggy (i.e., tall, thin and weak)—if they grow at all. I've found that even when grown next to a sunny south-facing window, most plants tend to struggle. For best results, you should supplement natural light with grow lights. If you don't, you're probably going to be disappointed.

There are several types of grow lights to choose from, ranging in both cost and effectiveness. As you read about the pros and cons of each below, keep in mind that there is more to a grow light than just the bulb. Unless you purchase a complete kit, you'll need to also find compatible reflectors, cords, ballasts and other parts.

Fig. Photo by Dineshraj Goomany

A quick note about light color: The whiteness of a light's output is measured in Kelvin units (K). The higher the degree of Kelvin, the "cooler" (bluer) it appears, and the lower, the "warmer" (redder) it appears. Blue light encourages compact, bushy vegetative growth. Red light initiates the flowering stage. Lights between 4,000 and 6,000K are known as "full-spectrum" lights because they produce both cool and warm light colors, much like natural sunlight.

Incandescent

Not much about the incandescent light bulb has changed since Thomas Edison invented it in 1879. While inexpensive and readily available, incandescent lights are the least efficient option for indoor gardening. They'll increase your utility bill, produce more heat than light, and burn out more quickly than other options. I wouldn't recommend using these to grow plants.

Fluorescent

A common—and perhaps the most popular—choice for home gardeners, fluorescent lights use ¼ the energy, last about 10x longer, and produce more light compared to incandescent lights (energy.gov). They're great for starting seeds and growing greens, herbs and other plants with low to medium light requirements.

The primary drawback with fluorescent lights is they don't produce enough light to effectively grow fruit-bearing plants, such as squash and tomatoes.

There are a few different kinds of fluorescent lights commonly used for indoor gardening:

- T5, T8 and T12 bulbs are long and tube-shaped. T5 is the narrowest and most efficient of these. Because of its smaller surface area, a T5 bulb produces the most intense light with the least energy. Plus, it doesn't produce much heat, so it can be positioned very close to plants to maximize its effect.
- Compact fluorescent lights (CFLs) are the same size as a typical incandescent light—but much more efficient. CFLs are useful for

lighting individual plants or seedlings, but not necessarily an entire Tower Garden.

The Tower Garden Grow Lights Kit uses high output T5 fluorescent lights. And in my own experience, they have worked amazingly well.

High-intensity Discharge (HID)

HID lights are typically used for larger growing spaces, such as commercial greenhouses. Producing a greater intensity of light, they're more efficient than fluorescent lights. But they're also large and expensive, which makes them less popular among home gardeners.

There are 2 kinds of HID lights commonly used for indoor gardening:

- Metal halide (MH) lights produce cool colored light (which, remember, is best for compact, leafy growth).
- High-pressure sodium (HPS) lights emit warm colored light. But using HPS lights alone often causes plants to become leggy.

If you decide to use HID grow lights, pair MH with HPS to encourage balanced growth for all types of plants.

Light-emitting Diode (LED)

Relatively new in the world of indoor gardening, LED grow lights are incredibly light, compact and efficient. LEDs can be configured to emit only red and blue light (the only colors of light plants can absorb), so they don't waste any energy. But, like HID lights, these will cost a pretty penny up front compared to fluorescent lights.

Tower Garden Support Cage

Once you get grow lights, you'll need a way to attach them to your Tower Garden. A standard Tower Garden Support Cage provides adequate support for LED and T5 grow lights.

Timer

A timer isn't required, but it will make your life easier. For healthy development, plants need sleep (i.e., darkness), just like we do. Using a timer to automatically turn your grow lights on and off will save you the trouble of remembering to do it. The Tower Garden Grow Lights Kit includes a timer.

Fan

It's important to facilitate air circulation around your plants, and running a small fan is an easy way to do this. Keeping the air moving will help prevent problems, such as leaf fungi and garden pests.

Rubber-backed Rug

Credit goes to Tower Gardener Nancy Kroupa for suggesting this one on the Tower Garden Facebook page. Place your indoor Tower Garden on a rubber-backed rug or mat to protect your floor from any accidental spills that may occur when refilling the water reservoir or performing other maintenance tasks. Like the timer, this is not necessarily required, but useful.

BEST PLANTS TO GROW INSIDE

When gardening indoors, I highly recommend starting your own plants from seed—don't use outdoor transplants. Why? Because when you bring a plant inside, there's a good chance you'll bring something (i.e., a pest or plant disease) with it. Even plants that appear healthy aren't worth the risk. Once garden pests take root (no pun intended) in your home, they can be difficult to eradicate, as there are no natural predators indoors.

Tower Tip: Grow pest-repelling plants, such as marigolds, indoors to help keep your Tower Garden problem-free! So what seeds should you start with? For most indoor gardening setups, greens, herbs and other leafy plants grow best. Here's a list of edible plants commonly grown indoors:

You're most likely to find success growing these plants, but that doesn't mean you can't try growing things like peppers or peas. Experimenting is part of what makes gardening so fun! But keep in mind, fruit-bearing plants aren't going to perform as well indoors as they would out.

Vegetables:	**Herbs:**
• Arugula	• Basil
• Broccoli	• Catnip
• Broccoli Raab	• Chamomile
• Brussels Sprouts	• Chives
• Cabbage	• Cilantro
• Chinese Cabbage	• Cumin
• Cauliflower	• Dill
• Chard	• Garlic Chives
• Collards	• Lavender
• Endive	• Lemongrass
• Escarole	• Marjoram
• Kale	• Mint
• Leeks	• Oregano
• Lettuce	• Rosemary
• Mesclun Varieties	• Sage
• Mustard Greens	• Stevia
• Pak Choi	• Thyme
• Radicchio	
• Sorrel	
• Spinach	

And they'll require additional attention—particularly when it comes to pollination. Without exposure to bees, wind and other natural methods of pollination, your plants won't produce fruit without your help. To learn more about manual pollination, read these tips.

IDEAL CONDITIONS FOR INDOOR GARDENING

Creating the right conditions for a successful indoor Tower Garden is easy. In addition to typical maintenance tasks, there are just a few important variables to manage:

Light

Running your lights for at least 14 hours a day should yield good results. But you can experiment with longer durations, as long as your plants still get 6+ hours of darkness.

Pump Timer Setting

When growing outside, you should set your pump timer to 15 minutes on and 15 minutes off. But in a cooler, climate-controlled environment, plant roots don't dry out as quickly as they do outside. Changing your timer to 15 minutes on and 45 minutes off when growing indoors is a way of both saving a little energy and preventing over-watering.

Temperature

Most plants will grow best between 65-75°F.

NOW GET GROWING

Congratulations—with this knowledge, you're ready to grow your own food indoors all year long! Season, light and space hindrances are a thing of the past. The easiest way to get started? Order a Tower Garden Grow Lights Kit. It comes with everything you need, so you can quickly assemble it, plug it in, and grow! Just be sure you're ready for all the fresh produce. I have had to get creative with my abundance of kale!

LIGHTING INDOOR HOUSEPLANTS

Houseplants are popular indoor decorations. Attractive and constantly changing, they add a softness of line and provide a bit of nature indoors. However, the ideal location of a plant for decoration may not be the ideal spot for plant growth. Lack of adequate light is the most common factor limiting the growth of plants in many areas of the home. Supplementary electric lighting is usually the easiest and least expensive way to provide enough light for plants that do not receive adequate natural light.

Why do plants need light?

Light provides the energy plants need to make the food required for them to grow and flower. Plants are the only organisms able to use light to produce sugars, starches and other substances needed by them as well as by other living organisms.

Is light color important to plants?

Certain colors in light rays are important for proper plant growth. Leaves reflect and derive little energy from many of the yellow and green rays of the visible spectrum. Yet the red and blue parts of the light spectrum are the most important energy sources for plants, and plants require more rays from the red range than from the blue.

Plants growing outdoors, in greenhouses or close to windows are exposed to a natural balance of the blue and red light rays that plants need. Where plants receive little or no natural light, you must provide additional light from artificial sources.

Which types of lights are best?

As a single light source for plants, incandescent light bulbs are not particularly good. They are a good source of red rays but a poor source of blue. They produce too much heat for most plants and, if used, must be located some distance from the plants, thus reducing the intensity of the light the plants receive. They are also about one-third as efficient as fluorescent tubes in converting electrical energy to light. Furthermore, a standard incandescent bulb's life is often only about 1,000 hours, whereas a fluorescent tube's life is normally 10,000 hours or more.

Fluorescent tubes provide one of the best artificial light sources available for plants in the home. Other light sources such as sodium-vapor and metal halide lamps may be used but are not as readily available or adaptable for home use. Fluorescent tubes are made in many sizes and shapes: circular, U-shaped, square or straight. Straight tubes in 2-, 4- or 8-foot lengths are used most frequently.

What is the best balance of artificial light?

Many indoor gardeners use cool-white fluorescent tubes. Warm-white fluorescent tubes also seem fairly effective, but fluorescent tubes listed as white or daylight are less desirable for indoor plant growth. Cool-white tubes produce a small amount of red rays in addition to orange, yellow-green and blue rays. However, the red light produced usually is not enough for plants unless windows or other artificial lights produce additional red rays.

A few incandescent bulbs in the growing area can furnish needed red rays. A general ratio of incandescent to fluorescent light is about 3 to 10, so for every 100 watts of fluorescent light, you should provide about 30 watts of incandescent light for a better red-to-blue light balance. Special fluorescent tubes also have been developed for growing plants. These have a higher output in the red range to balance the blue output. Many home gardeners have found that these tubes can be used in combination with cool-white tubes. Use one special plant-growing tube to each one or two cool-white tubes. This method is more economical than using all special tubes, since cool-white tubes cost less than the special plant-growing tubes. Also, fluorescent plant-growing tubes use less electricity and produce less heat than incandescent bulbs, and you will not have to provide fixtures for both incandescent bulbs and fluorescent tubes.

May I use spotlights or other special bulbs?

Yes, although they are less effective than fluorescent lights and the combinations described previously. However, fluorescent fixtures may not be suitable for some locations. Reserve these special light sources for situations where supplementary light is essential. Self-reflectorized spot lamps coated to emit more blue light are now available.

How much light should plants receive?

The amount of light necessary varies with each plant. In general, the light fixtures available for home plant lighting make it practically impossible to produce too much light for most plants.

Plants that can adapt to interior settings usually are divided into three general categories: those suitable for low, medium and high light intensities. The categories generally indicate the minimum light required. Growth is often best at the higher end of these suggested light ranges.

Low-light plants

Plants referred to as low light intensity plants generally should receive between 50 and 250 foot-candles. (A foot-candle is a measure of light intensity. A foot-candle is defined as the amount of light received by 1 square foot of a surface that is 1 foot from a point source of light equivalent to one candle of a certain type.) Under artificial light, a few plants in this group can be maintained at as little as 10 foot-candles. One way you can estimate the amount of light available is to calculate the number of watts available per square foot of plant area. Low-light plants should receive between 10 and 15 watts of fluorescent light per square foot of growing space. A single fluorescent tube, such as a 2-foot 20-watt tube or a 4-foot 40-watt tube without any other light provides only enough light for plants in this category.

Medium-light plants

These plants prefer 250 to 1,000 foot-candles. Best growth occurs above 750 foot-candles unless plants also receive extended periods of direct sunlight. Give them artificial light in the 500 to 1,000 foot-candle range or 15 or more watts per square foot of growing area. While plants in this group can be held in the 250 to 500 foot-candle range, growth is best with more light. A fixture containing two fluorescent tubes is sufficient for plants in the low to medium light range. Adjustments in the number of tubes used may be made if you regulate the distance between the tubes and plants.

High-light plants

These plants generally are less satisfactory for growing under artificial lights in the home. However, if you want to try, use special high-intensity lamps.

These plants need at least 1,000 foot-candles, or 20 watts per square foot of growing area, but should have higher intensities for best growth and flowering. Fixtures containing three to four fluorescent tubes are necessary for plants requiring high light.

Table. Light output (in foot-candles) measured at various distances below fluorescent lamps. Output in parentheses is measured 1 foot on either side of a line directly below the lamps. All lamps are standard 40-watt tubes.

Distance	Type of fixture		

0.5 foot	500	700	900
1 foot	260 (200)	400 (260)	600
2 feet	110 (100)	180 (150)	330
3 feet	60 (60)	100 (90)	
4 feet	40	60	100

How far from the light should I place plants?

Most plants should be located with the tips of the plants 6 to 12 inches from the light source. The intensity of light drops rapidly as the distance from the light bulbs or tubes increases. This reduction of light intensity with distance below and to the side of tubes. Fluorescent tubes also do not produce as much light at the ends as they do in the center. Therefore, the brightest spot under a fluorescent fixture is directly beneath the center of the tubes.

The light fixture's position should be adjustable so you can keep the distance between the light and the plant fairly constant. Fluorescent shop or workroom fixtures often are hung on chains with S-hooks for easy adjustment. These fixtures are easily raised or lowered from link to link. If the fixture is not movable, you may make some adjustment by raising plants on stands, shelves or boxes.

How long should I use lights?

In most cases, plants receiving no outdoor light should be lit from 16 to 18 hours each day. If some additional light is received, 12 to 14 hours each day may be adequate. Lights should be used at the same time that plants receive window light. Using lights at the beginning or end of the day will not usually be as effective as using lights during daylight unless natural daylight is quite bright.

How can I get the most from artificial light?

Reflectors and reflective surfaces can maximize the available light. Bulbs with self-contained reflectors are helpful.

Porcelain-coated reflectors are excellent and require little maintenance. Keep reflectors clean and free of rust or any coating that reduces their effectiveness. White paint or aluminum foil beneath or around the growing area helps reflect light and makes it more efficient. Space plants far enough apart to allow light between them. Arrange plants so they do not shade each other. Keep tubes clean and replace old tubes promptly.

How should I light the bottom of a tall plant?

You may want to supplement light placed above the plant with spotlights around the base of the plant and directed on the lower leaves. You can also use fluorescent tubes in a vertical position to provide side lighting from the top to the bottom of a plant. This vertical position also can be used for smaller plants arranged on shelves.

Do I really need to measure light?

The eye is a poor judge of light intensity because it automatically adjusts to different light levels. Light measurements are helpful in setting up a plant growing area but should be used only as a guide.

Light meters that measure foot-candles may be used if available. Photographic light meters normally do not read in foot-candles, but some manufacturers supply a conversion table. When available, these may also be used. Calculating wattage per square foot of growing area is a useful and easy way to estimate light required. With this approach, light measurements are unnecessary unless problems develop.

How can I tell if there is enough light?

The growth pattern of the plant can be a good indication. No growth may indicate poor light but could be a sign of other problems as well. A healthy plant under poor light may develop long internodes (length of stem between leaves). Leaves may develop smaller than normal on some plants. Color of many plants in poor light may be pale green, and lower leaves may yellow and drop.

What window produces the best light?

Consider size, direction, overhang and shade from trees or buildings. Large windows provide the best growing conditions and allow plants to be placed fairly far back into a room. But even low-light plants usually do not receive enough light at distances greater than 10 feet from an average window. The best windows for plants are those not shaded by a large overhang, trees or structures.

Windows facing south provide the brightest light conditions for the longest duration. In winter, any houseplant benefits from the light of a south window. However, plants that do not need bright light may be sunburned by the bright light at south windows in late spring, summer or early fall. Place plants requiring less light, such as African violets, at a north window or to the side or interior of a large south window during these times. South windows are most appropriate for plants requiring bright light and some direct sunlight.

East and west windows are well suited to many plants in the medium light range, while north windows are satisfactory only for plants requiring the lower light levels. These plants should not receive direct sunlight.

Can lights keep plants from blooming?

Some plants, generally known as short-day plants, can be kept from flowering under the light durations normally used for artificial lighting. Best known in this category are the poinsettia and chrysanthemum. To induce flowering indoors, give these plants only about 10 hours of light each day until flowers become visible and color shows.

Will artificial lights start seeds?

Vegetable, annual flower and some perennial flower seeds may be started successfully indoors under lights for later planting in the garden. For stocky growth, place seedlings within a few inches of the tubes as soon as germination begins.

Is a light timer necessary?

A timer is a valuable asset because lights should be turned on and off regularly and consistently. Twenty-four-hour timers available from electrical supply houses are adequate. The electrical cord from the timer should be the three-prong type, or you should use a grounded adapter. The use of water around plants makes grounding electrical fixtures important.

Plants for indoor lighting

African violet

- Saintpaulia species
- This is one of the most satisfactory flowering plants for growing under lights; it grows and flowers well between 500 and 1,000 foot-candles. Light 16 to 18 hours each day.

Aluminum plant

- Pilea cadierei
- This small plant with colorful leaves tolerates a low light level but needs high humidity. Other suitable related plants include artillery fern and moon valley pilea.

Arrowhead vine

- Syngonium podophyllum
- These small plants are bushy and later, creeping or climbing. They may be kept at a low light level, but they are denser at a medium level.

Asparagus ferns

- Asparagus densiflorus 'Sprengeri'
- Asparagus setaceus (formerly plumosus)
- These durable potted plants with fleshy roots produce fernlike leaves. Although tolerant of low light, the plants grow more abundantly and densely in a medium light range.

Aucuba-leaf croton

- Codiaeum 'Aucubaefolium'
- Crotons are not appropriate for the light garden as they require high light for good color development and growth. However, this smaller type with yellow spots can be grown at medium light intensities.

Begonias

- Angelwing, B. coccinea
- Beefsteak, B. xerythrophylla
- Iron cross, B. masoniana
- Flowering, B. semperflorens
- Painted B. rex varieties
- Most plants of this family are suitable for growing in the medium light range. Begonias like high humidity and constantly moist soil. Give them a long light duration, from 14 to 18 hours each day.

Cacti and succulents

- Many plants may be grouped in this general category. These plants need high light intensities; most are not satisfactory for growing under artificial lights. However, you can use lights to maintain them for limited periods when they cannot be exposed to bright light or direct sunlight. When artificial lights are used, provide high intensities, and keep plants close to the source.

Cast iron plant

- Aspidistra elatior
- This plant tolerates cool locations and poor light; it is one of the best plants where only low light conditions exist. A variegated form is also available.

Chinese evergreen

- Aglaonema modestum
- Aglaonemas are relatively small tropical plants but can tolerate light as low as 10 foot-candles. Other species with different leaf form and variegation are also available.

Cissus species

- Grape ivy, Cissus rhombifolia
- Kangaroo vine, Cissus antarctica
- Cissus are climbing plants that prefer medium light conditions for best growth, although kangaroo vine tolerates lower light. Although

better for greenhouse culture, another beautiful relative is the rex begonia vine, Cissus discolor.

Dieffenbachia

- Dieffenbachia maculata
- Many satisfactory species and varieties are available. Plants have large, showy leaves with spotting and variegation. Medium light range is best and will prevent loss of lower leaves. Overly tall plants can be cut back. Dieffenbachia benefits from supplementary lighting in poorly lit areas.

Dracaena species

- Corn plant, Dracaena fragrans massangeana
- Dragon tree, Dracaena marginata
- Janet Craig dracaena, Dracaena deremensis 'Janet Craig'
- Ti plant, Cordyline terminalis 'Ti'
- Most dracaenas become large plants. They are well adapted to house conditions and tolerate low light although they also do well in medium light. They need to be kept warm and constantly moist for best growth. Ti plant develops best leaf color in bright light and therefore is less suitable for locations where it receives only artificial light.

Ferns

- Birdsnest fern, Asplenium nidus
- Boston fern, Nephrolepis exalta bostoniensis
- Holly fern, Cyrtomium falcatum
- Maidenhair fern, Adiantum species
- The word "fern" brings to mind the Boston, or sword, fern with its many varieties. These ferns are durable houseplants that tolerate low light but prefer a medium light level. Keep them cool and moist. Holly fern likes cooler house temperatures; however, birdsnest fern prefers warmer temperatures. Ferns may develop brown leaves or leaflets at low humidity. Maidenhair ferns should never be allowed to become dry.

Ficus species

- Rubber plant, Ficus elastica varieties
- Fiddleleaf fig, Ficus lyrata
- Weeping fig, Ficus benjamina
- These large plants are well suited to most home conditions and are best grown in a medium light range. Low light may result in a drop of foliage. Keep soil evenly moist. They tolerate low humidity.

Flame violet

- Episcia cupreata varieties
- These close relatives of the African violet have beautiful foliage in many color patterns. Flowers are red, pink, purple and yellow. Light

needs are the same as for African violets. The plants need high humidity and temperatures that do not drop below 65 degrees Fahrenheit.

Gloxinia

- Sinningia speciosa fifyana varieties
- This colorful African violet relative needs a medium light range to prevent elongation of the stems as well as to promote abundant flowering. Buds sometimes appear but fail to develop. This may indicate insufficient light, too low humidity or mite damage.

Herbs

- Culinary herbs can be grown indoors, but for best growth, most of these require fairly bright light. If you attempt them under artificial light, use as much light as possible with a long duration.

Hoya

- Hindu rope plant, Hoya carnosa 'Hummels compacta'
- Wax plant, Hoya carnosa.
- Hoyas are vining plants with thick, waxy leaves. They tolerate low light but need medium light to grow and high light to flower. Plants should not be moved; allow them to remain undisturbed. Keep slightly cooler and drier in winter.

Ivy

- Hedera helix varieties
- These vining plants with leathery leaves grow best in medium to bright light areas. They need constant moisture to prevent leaf drop, and they tend to grow better when kept slightly cool indoors.

Jade plant

- Crassula argentea
- This succulent plant is well adapted to indoor conditions. Although most crassulas prefer sun, jade plant will tolerate lighting as low as 25 foot-candles; however, growth will be thin and leaves small. Avoid overwatering and tight soils.

Little-leaf schefflera

- Schefflera arboricola
- This medium-sized plant has waxy leaves that are palmately compound. It has a bushy type of growth and, although it prefers higher amounts of light, it will tolerate medium light quite well.

Norfolk Island pine

- Araucaria heterophylla (formerly excelsa)
- This is a durable plant for home conditions. Although tolerant of poor light, it develops better growth at medium light conditions. With poor care and low light, it may lose symmetry.

Orange

- Dwarf or calamondin, Citrus mitis
- Potted citrus plants prefer bright light for best growth and fruiting. You may keep them at medium light intensities during midwinter. The calamondin orange is the most popular potted citrus, although Meyer lemon and Ponderosa lemon are also satisfactory for home use.

Orchids

- Few orchid species are well suited to the average home's artificial light conditions. Seedling plants do well under artificial lights; however, mature plants need bright light for best growth and flowering. Some genera better adapted to the light garden include Epidendrum, Paphiopedilum, Brassavola and Phalaenopsis. All orchids need excellent drainage and air movement.

Palms

- Areca palm, Crysalidocarpus lutenscens
- Dwarf date palm, Phoenix roebelenii
- Neanthe bella palm, Chamaedorea elegans 'bella'
- Bamboo palm, Chamaedorea erumpens
- Kentia palm, Howeia forsteriana
- Neanthe bella and kentia palms are most tolerant of low light conditions. Other palms are best suited to medium light areas. Bright light may cause fading in some species. Palms prefer fairly uniform moisture, but occasional slight drying can be beneficial. Palms grow relatively slowly indoors. Keep close watch for spider mite damage.

Peperomias

- Baby rubber plant, Peperomia obtusifolia
- Emerald ripple, Peperomia caperata
- Many other suitable species
- Most peperomias have fleshy leaves and stems. They are small plants well adapted to home conditions. Provide a medium light level, although peperomias can tolerate poor light for limited periods. Avoid overwatering, which may promote stem rots.

Philodendrons

- Heartleaf philodendron, Philodendron oxycardium
- Splitleaf philodendron, Philodendron pertusum
- Other suitable species
- The heartleaf philodendron tolerates low light; most prefer a medium light range for best growth indoors. Most are well adapted to home growing. They should be maintained at fairly uniform moisture.

Pilea species

- These small but colorful plants (e.g., Pilea caperata, Emerald ripple pilea) are excellent choices for table tops, dish gardens and terrariums. They require the same light conditions as do the peperomias.

Pleomeles

- Lance dracaena, Pleomele reflexa
- Malaysian dracaena, Pleomele thaliodes
- These plants are gaining popularity because of their durability as potted plants and tolerance to low light conditions. They grow best in a medium light range with uniform soil moisture.

Pothos, devil's ivy

- Epiprenum aureus
- This popular variegated climbing vine is similar to heartleaf philodendron and satisfactory for low and medium light conditions. It requires less uniform moisture than philodendron.

Sansevieria, snake plant, mother-in-law-tongue

- Sanseveria trifasciata
- This very popular and durable plant grows well indoors. It tolerates poor light but makes best growth in medium to bright light conditions. Dwarf and variegated forms are available. Avoid overwatering.

Schefflera, umbrella tree

- Brassaia actinophylla
- This plant prefers bright light but tolerates medium light or even low light for limited periods. It will drop foliage in extended periods of poor light or if kept too wet or too cool. Spider mites are a common problem.

Spathiphyllum, peace lily

- Spathiphyllum sp., e.g., 'Supreme' or 'Sensation'
- These plants adapt to low light conditions but need medium light to flower. Their white, anthurium-like flowers are long lasting. Do not allow plants to dry out.

Spider plant, airplane plant

- Chlorophytum comosum 'Vittatum'
- Long drooping leaves and runners producing small plants make this a popular hanging plant. Solid green and variegated types are available. They tolerate low light but grow best in the medium range. Keep well watered.

Swedish ivy

- Plectranthus australis
- This creeping plant is used for hanging containers. Although it tolerates low light, more dense growth and branching occur at medium light intensities.

INDOOR LIGHTING FOR PLANTS

Artificial lighting from light bulbs indoors can be used to start seedlings in spring, provide supplemental light for sunlight to many plants, and to provide the sole source of light for low to medium light plants. The main aspects to consider when providing light for plants indoors is the quality, quantity, and duration of the light.

Quality refers to the actual wavelengths the lights provide to the plants. Light may look white to us, but is in reality made of many different wavelengths as seen in rainbows or when light strikes a prism. It is the red and blue parts that the plant uses for energy and growth, so these need to be provided by indoor lights. Terms you may see for light bulbs that provide these are "natural", "full spectrum", or "balanced." They cost more than the usual incandescent bulbs that mainly provide red light to plants, or the cool white tubes that mainly provide blue light.

A less expensive solution for a balanced light quality is to use both incandescent and cool white lights, or cool and warm (appearance, not temperature) white tubes in fluorescent fixtures. Keep in mind incandescent bulbs only may last 1,000 hours compared to 10,000 hours for fluorescent. If using both incandescent and fluorescent, a balance ratio of 3 to 10 is best. So for every 30 watts from an incandescent bulb, provide 100 watts from fluorescent.

Additional terms you will see when looking at fluorescent tubes are "T" numbers. These refer to the diameter, in eighths of an inch, of tubes. So a T8 tube is eight eighths, or one inch, in diameter. Older tubes are T12, with most newer ones T8 or even T5. These newer ones tend to be much more energy efficient, and even with lower wattage can produce more light due to new technology and materials in their production. There even are high-output tubes with longer life, so even though more expensive at first they are cheaper in the end.

Quantity of light is crucial and, in addition to type and wattage of bulb, is adjusted by distance of lights to plants. Fluorescent tubes give off little heat, so can be placed as close as four to six inches above seedlings and plants. Incandescent and similar bulbs give off more heat, so need to be kept a foot or more above plants. If you see browning or "burning" of leaves and leaf edges, this may be a sign your light is too close to plants.

Most plant growing fixtures you find in stores and catalogs utilize fluorescent tubes. These growing stands can be quite attractive, and expensive. If aesthetics aren't crucial, such as in a heated basement area, you might consider making your own fixtures from lumber. If making your own, consider bolting them together in case they need to be taken apart to move or store. Use adjustable chains for hanging fixtures so they can be placed at the proper height, and moved up as plants grow. For seedlings, and the most light on other plants, use either a three or four-tube fluorescent fixture, or two two-tube ones. When using any bulbs, the reflector can make a difference how much light the plant receives. Reflectors are advised for most indoor lights, especially tube fixtures. Light quantity can drop off dramatically, by two fold or more, near edges of reflectors. Seedlings and plants growing at an angle toward brighter light is a sign the quantity is not sufficient over them. I make sure, when starting seedlings, to have the fluorescent light reflectors out past the edges of the seedling flats. You can line older reflectors with shiny aluminum foil to provide more light, or even place such foil or white surfaces behind plant fixtures for reflection.

Light quantity also drops dramatically near the ends of fluorescent tubes, even new ones. If you have older tubes, and they are getting a bit dark near the ends of the bulbs, this is a sign they should be replaced. If you plan to grow many plants indoors under lights, you may want to look online, at a complete garden store, or at a camera shop for an inexpensive light meter. This can be revealing to see how much light is provided, and where, for your plants and even for reading lamps.

Light for growing is commonly measured in foot-candles—the amount of light given off by a candle one foot away. Two common T8 tubes placed six inches above plants will provide about 700 foot-candles. Reading lights, for comparison, often provide about 50 foot-candles.

Low light plants such as Chinese evergreen and Peace lily need between 50 and 250 foot-candles. Medium light plants such as African violets, begonias, dracaena, dumb cane, flame violet and seedlings need 250 to 1,000. High light plants such as most herbs and orchids need over 1,000 foot-candles.

Many plants may tolerate one level but prefer another. English ivy, peperomia, philodendron, and most ferns tolerate low light but prefer medium (in addition to high humidity for ferns). Many high light plants will tolerate medium light, but may be smaller, with smaller leaves, or may not flower and fruit if this is their attraction.

Light quantity and light duration are related, as a lower level of light often can be made up with a longer duration. If growing seedlings indoors, 16 to 18 hours of light at day is recommended, at 65 to 75 degrees F. This duration, or even 12 to 14 hours with brighter light, should be used if the sole light source for low to medium light plants. If using indoor lighting merely to supplement natural light from windows, you only may need a few hours additional at the end of the day. Don't keep lights on continuously as plants need periods of darkness, with 55 to 65 degrees F ideal.

Natural and supplemental light is best for high light plants, as it is hard to get the quantity they need indoors in most homes solely from light bulbs. There are high-intensity fixtures for high light plants, as seen in sports arenas, that are quite expensive, require much power, and are best for serious hobbyists or commercial growers. Various inexpensive timers to control light fixtures are available at most hardware stores. Some multi-plug power strips even come with timers. Other items to consider with artificial light include heating mats. These especially are important if starting seedlings. Trays to prevent water runoff are needed for potted plants. These can be lined with gravel and kept moist in order to increase humidity around plants to a minimum 50 percent or so.

Artificial lighting is useful for starting seedlings indoors, either to get a jump on the season or for those such as pansies that take many weeks to germinate and grow. Make sure before placing outdoors in late spring such seedlings, or even potted plants that have been solely or primarily under artificial lights, that you adjust them gradually over a couple of weeks or more to the "real" and higher light of sun. You may just want to leave potted plants in part to full shade. Put seedling flats here too, or cover with lightweight fabrics as used for frost protection.

If you haven't tried growing plants under lights, join the roughly 2.7 million households that buy indoor plant lights each year, and the estimated 10 to 15 million households that grow plants under lights indoors.

INDOOR PLANTS THAT PURIFY THE AIR AROUND YOU NATURALLY

Bringing potted plants into a living space to liven it up is a trick that's been used by interior designers for years, but did you know that our leafy friends are also powerful filters that purify the air around us? In fact, several studies have been conducted showing that certain plants can rid a room of up to 89 percent of harmful VOCs like formaldehyde and xylene. If you think about the prices of some of those fancy air filtration systems out there, it's a bit surprising that more of us don't just purchase some plants instead. If you or anyone in your family has allergies, smokes or just wants to breathe fresher, cleaner air in their homes, read on for 7 indoor plants that purify the air around you as well as which specific pollutant each one targets and removes.

Dendrobium and Phaeleonopsis Orchids

Orchids have a bad reputation as being finicky and difficult to grow, but really, the opposite is true. On our last trip to Costa Farms, we learned that orchids actually love to be neglected and most people end up killing their orchids with kindness (too much water and sunlight).

Aside from being easy to take care of, orchids rid the air of xylene—a pollutant found in many glues and paints—so they make wonderful housewarming gifts for anyone who recently moved into or renovated a new space. Unlike some other plants, orchids also respire and give off oxygen at night, so they're great for the bedroom.

Palms

The palm family of plants, also known as Arecaceae or Palmae, is extremely popular and it's easy to see why. These hardy houseplants are easy to grow and perfect for lifting people's spirits and distracting from otherwise drab surroundings, and they're also known to be natural air purifiers. Palms specifically target and remove formaldehyde, benzene, and carbon monoxide (which is especially helpful if someone in your household smokes cigarettes).

Peace Lilies

The peace lily, a.k.a. spathiphyllum, is an ideal plant to have in your home if you love flowers but don't want to buy bouquets that die after a few days. Spathiphyllum thrives in the shade in temperatures below 55 degrees F, and removes harmful toxins like acetone, ammonia, benzene, ethyl acetate, formaldehyde, methyl alcohol, trichloroethylene, and xylene.

Ferns

Ferns have a reputation for being a bit mundane but most people don't realize that they're actually fascinating plants that have survived since Prehistoric times!

They're favored for their soft, feathery leaves, and it's those same large fronds that help rid the air of pollutants like toulene and xylene, which are found in many paints, nail polishes, and glues.

Schefflera

Schefflera are easily recognizable because they have glossy, sturdy-looking oval leaves that almost look unreal because of their waxy shine. They're really hardy and long-lasting so they make great investment plants as long as you keep the leaves dust-free and wipe them down once in a while. In addition to looking great, they're also known to soak up nasty toxins like benzene, formaldehyde, and toluene so, like palms, they're good for households where there's a smoker.

Anthuriums

Anthuriums make lovely gifts because of their exotic-looking blooms, but they ain't just a pretty face! Their large, dark leaves suck up ammonia, formaldehyde, toluene and xylene, so they're a thoughtful present for a workplace (especially around copiers, printers, or adhesives).

Song of India (a.k.a. Dracaena reflexa)

As versatile as its name is poetic, Dracaena reflexa or "Song of India" is easy to identify because of its telltale green, lime, and yellow leaves. These plants are easy to grow in both high and low light and absorb undesirables like formaldehyde, toluene, and xylene.

Pothos

Pothos is characterized by its golden, heart-shaped leaves and is extremely popular in North America. It's a hardy plant that can survive in lower light and colder temps and is great for offices and homes since it rids the air of carbon monoxide and formaldehyde.

Massangeana Cane

The Massangeana plant may be hard to pronounce but it's easy to love. Native to Africa, it has a wild look that makes it ideal for decorating your home, and it also sucks formaldehyde from the air.

Philodendrons

Philodendrons are easy-care houseplants that need very little attention. Their unique coloring makes them an attractive addition to your home and they're known to ride the air of xylene.

Eager to learn even more about how to clean the air inside your home with plants? How to Grow Fresh Air is a wonderful resource with many more types of air-purifying foliage and comes highly recommended by Inhabitat's editor-in-chief, Jill Fehrenbacher. You can also visit Costa Farms' helpfulO2forYou website for more information.

4

Conservatory for House Plants

A BRIEF HISTORY OF HOUSE PLANTS

As far back as the time of the great pharaohs and the pyramids, the people of Ancient Egypt decorated their homes with plants. Displaying plants in containers was popular in Egyptian art. The Greeks and the Romans built their homes around a central atrium of containerized plants. And even though these plants were out in the open elements, this atrium was considered apart of the home. This was the fashion of such civilized societies.

During the age of European discovery, the explorers who went to the New World found many new tropical treasures, many of which did not survive Europe's climate, so they were brought indoors. For centuries curious explorers and botanists, called plant hunters, searched the tropics for new and exciting plants to bring back to Europe.

Since there was no suitable climate for these new treasures, plant houses needed to be constructed. The first of these was called an orangerie, like the one built at Versailles in France that housed the citrus and palms of KING LOUIS XIV over the winter. Later conservatories and greenhouses like we know them today were developed.

During the Victorian Age, exotic plants become extremely popular, but the transport of them was very difficult because of the lack of warmth, rooting media, and care on the ship ride home.

On their long journeys, plants would become diseased or damaged and die until 1833, when Dr. Nathaniel Ward developed a glass case to transport the plants. These Wardian Cases were self sufficient on long journeys and became a fad in Europe for transporting exotic plants. Today we call these cases terrariums.

Also during this Victorian age, transition was in the air and technology changed the lives of the people. There was less free time in men's lives and the role of women changed in the home. Besides just taking care of the home, women chose horticulture as a hobby to become more active. This was at first just outdoor gardening, but there was a shift to gardening indoors in poor weather and the use of houseplants that we know today came into existence. This hobby took off and became extremely popular with both men and women and soon almost everyone had houseplants, especially in the cities. Before the 1940's the home environment was not suitable for tropical and subtropical plants, the popular foliage plants of today. After this time though, more precise control of indoor temperatures in both summer and winter greatly improved and there was no longer a need for conservatories and greenhouses to keep these plants over the winter.

DESIGN IDEAS WITH HOUSE PLANTS

When designing a garden outdoors you are working in a setting that is more natural to the plant. This is not true when you are designing with houseplants.

The domestic situation is highly artificial, so where you place the plant in the home depends on their specific cultural requirements. Many times artificial lighting is required.

But also size, shape, texture, and color of the plant must complement your homes decor. When indoors, the room size, furniture, and windows are all in human proportions. When designing with houseplants you must keep the design in this proportion unlike the role of outdoor plants. It is easy to put too many large plants, or large groupings of plants in small areas. Plants are not the dominant feature in the home. They are meant to compliment the furniture and the existing design.

By using the specific cultural requirements of the individual plants, the information below, and other literature, one should be able to decorate their home or office with their own personnel taste in plants.

- Locations
  - Living Rooms
  - Kitchens and Bathrooms
  - Bedrooms
  - Halls and Stairwells
  - Open Plant Windows
  - Terrariums
  - Garden Rooms
  - Balconies and patios
- Display Supports
  - Tables
  - Hanging Containers
  - Wall Containers and Brackets
  - Other Vertical Supports and Containers
  - Pots, Troughs, and Tubs
  - Shelving

GREENHOUSE, CONSERVATORY AND HOUSEPLANTS

IN THE GREENHOUSE

Maintain a minimum of 5°C to prevent fuchsias, pelargoniums and other tender plants being killed by the cold. Higher temperatures (at least 12°C/54°F) will be needed for tropical plants.

Pot up any Begonia, Gloxinia and Achimenes(such as A. 'Show Off', left) that you started off earlier in the spring and are now large enough to re-pot. Don't forget to prick out seedlings before they get too crowded, and then to pot them on as individual transplants as soon as they are large enough.

Check if plants need watering at least every few days. Seedlings will need daily attention. Try growing on plug plants in your greenhouse. They are a relatively cheap source of large numbers of plants, and avoid the need for propagation facilities and time-consuming pricking out. Don't forget to give greenhouse plants more space as they put on new growth. This will help to prevent disease, and to contain early pest infestations.

Greenhouse bulbs that have finished flowering (e.g. freesias and Lachenalia) can now be fed weekly with high potassium feed (such as tomato fertiliser), to build up the bulbs before resting them over the summer. Reduce feeding and watering as the foliage starts to die back, and then lay the pots on their sides under the greenhouse bench until next autumn.

Remember that temperatures can still drop at night. Last year we had frosts until late May in southern England. Seedlings in particular object to wide temperature fluctuations. A heated propagator could be a cost effective answer if your greenhouse is otherwise full of hardy mature plants.

If the weather is warm and sunny, you may need to start damping down the floor of the greenhouse in the morning, splashing water over it to increase humidity levels in the whole glasshouse. This will help reduce problems with red spider mite which favour a dry environment.

However, very high humidity can encourage diseases such as grey mould so also be sure to open vents and doors of greenhouses during warmer days, ideally installing automative vent openers which will open gradually as the temperature inside the greenhouse increases. Keep alpine houses well ventilated too.

Apply a light covering of shade paint on greenhouse panes if weather turns warm at the end of the month, to prevent plants becoming scorched by bright sunshine. Alternatively, a sheet of newspaper or some shade netting from the garden centre work fine as temporary shading for seedlings and young plants. You could use spare space in the greenhouse to warm up grow-bags ready for planting.

Houseplants

Increase the watering of indoor and conservatory plants as days lengthen. Check if plants need watering at least every few days.

Start using liquid feeds once plants show signs of growth. Begin feeding citrus plants with a proprietary citrus feed, or using a high-nitrogen feed with added trace elements. Pot up houseplants showing signs of being root-bound or top dress large containers with fresh compost. Clean shiny-leaved plants with a damp cloth and spiky plants with a soft brush.

Cool conditions and regular watering will help keep potted indoor azaleas looking good for longer. Remember to water azaleas with rainwater rather than tap water if you live in a hard water area.

Deadhead Hippeastrum (amaryllis), leaving the flower stalks to die down naturally. Keep feeding and watering, and you may be treated to further flowers in August, as well as the normal blooms next winter. Alternatively, feed only for a few weeks (to build up the bulb) and then reduce the watering, let the leaves die back, and allow the plants to dry out completely. The bulbs can be 'rested' like this under the greenhouse bench or in a cupboard if they are houseplants), to be brought out in late summer or autumn, for flowering the following winter.

Rest pot cyclamen that flowered over the winter, reducing watering until the foliage starts to die down. The pots can be placed under the greenhouse bench (or in cupboard if they are houseplants), to be started back into growth in late summer. Brush up fallen compost and debris, and pick off dead leaves from plants. This will help prevent pests and disease spreading.

Propagation

Take leaf cuttings of streptocarpus plants. Hydrangeas and fuchsias can be propagated from softwood stem tip cuttings.

Pest and disease watch

Brush up fallen compost and debris, and pick off dead leaves from plants. This will help prevent pests and disease spreading. Damping off of seedlings can be a problem with sowings under glass. Keeping all your equipment clean and always using fresh seed compost will help prevent this. Insects start to emerge as temperatures increase. A mild spring can see pest problems developing earlier than usual.

PLANTS FOR THE WARM CONSERVATORY

If your conservatory is to be used a a living area, then maintaining a steady temperature of 60 - 70 degrees F (15.5 - 21 degrees C) is acceptable for most people. This is classed as a warm conservatory and also makes an ideal home for many tropical and sub-tropical plants. If your conservatory is to be used a a living area, then maintaining a steady temperature of 60 - 70 degrees F

(15.5 - 21 degrees C) is acceptable for most people. This is classed as a warm conservatory and also makes an ideal home for many tropical and sub-tropical plants.

- Columnea (Goldfish Plant) Yellow, orange or red tubular flowers.
- Ideal hanging plant.
- Gardenia (Cape Jasmine) Strongly scented, waxy white flowers
- from March - July, with dark green, glossy leaves.
- Hibiscus (Rose of China) Large papery flowers from Spring to
- Autumn and dark green glossy leaves.
- Saintpaulia (African Violet) Oval, hairy, green leaves.

- Produces several flushes of flowers each year in good light.
- Stephanotis (Madagascar Jasmine) Heavily scented, waxy
- white tubular blooms and glossy oval leaves.
- Tillandsia (Air Plant) Unusual group of plants forming rosette
- leaf clumps. Daily misting is all the care required.

Plants for The Cool Conservatory

The cool conservatory has a minimum winter night temperature of 40 - 45F (4.5 - 7 C). This is suitable for a wide range of tender plants.
Flowering Plants

- Bulbs A good variety available for Spring colour; Crocus, Narcissi, Tulips &
- Snowdrops to name but a few.
- Chrysanthemum An abundance of showy flowers in Summer above mid-green
- Foliage.
- Cyclamen Attractive foliage with swept-back flowers on long stalks. Some
- are sweetly perfumed.
- Jasminum Climbing plant that bears flowers in clusters on twining stems with
- Much divided leaves.
- Miniature Rose Bears lovely 1-3cm blooms in a variety of colours, many
- Are fragrant.
- Solanum - (Winter Cherry) Orange or red berries, amongst dark green
- Leaves. Very festive!

Foliage Plants

- Cacti Require a well lit spot, especially in Winter. Available in a variety of shapes and sizes.
- Fern (Pteris) Easy to grow, producing handsome fronds in a variety leaf shapes.
- Hedera Helix (Ivy) Useful trailing or hanging plant.
- Yucca (Cane Tree) Requires a well lit spot in Winter. Bears a crown of leathery, green leaves on a woody trunk.

Plants for The Intermediate Conservatory

With a minimum winter temperature of 50F (10C), you would have an intermediate conservatory, which suits a wide range of plants. Intermediate and cool conservatories are, of course, rather chilly when maintained at the minimum temperatures, but remember, they will often be warmer than this as the sun will raise the temperature.

Flowering Plants

- Bougainvillea (Paper flower) Lush and shrubby, leaf shedding, climber with pink/white flowers.
- Bromeliads Related to the pineapple. Colourful and bizzare bringing
- a touch of the tropical rain forest.
- Campanula (Star of Bethlehem) White and blue flowers. Trailing and groundcover.
- Citrus (Oranges, Lemons) Evergreen shrub or tree with highly fragrant flowers in
- Spring/Summer.
- Cymbidium Evergreen Orchid Flowers February to June (named varieties).

- Streptocarpos (Cape Primrose) Trumpet shaped flowers, available in white, blue,
- purple, pink and red.
- Thunbergia (Black Eyed Susan) Orange flowers with dark eye. Annual climber or
- ideal in hanging baskets.

Foliage Plants

- Hedera (Gloire de Marengo) Attractive and colourful with large leaves.
- Rhocissus Rhomboidifolia (Grape Ivy) Evergreen climber.
- Tradescantia (Wandering Jew) Evergreen trailer available with purple or green variegated leaves.

Containers

Many plants, especially trailing kinds, are effectively grown in elevated containers like hanging baskets, enhancing the roof area with a splash of colour. Choose the moulded plastic type with build in drip tray for less mess when watering.

Half baskets and wall pots that are fixed directly to walls, are also ideal for trailers. The back wall of a conservatory offers an ideal location for climbing plants, although they will require additional support. For displaying plants on the floor, there is a huge choice of decorative plastic, stone and terracotta pots available in many styles and sizes. Group several pots together when displaying to create an impact.

Also available are self watering tubs and planters, which are especially useful over periods away from home! Throughs are ideal for low growing plants such as Impatiens (busy lizzie) and Begonias, whilst Passion flowers do best in deep containers as they tend to be deep rooted.

Plant Support & Training

Offering plants a system of support is really straight forward. Canes and moss poles are ideal for foliage plants using twist or ring ties to secure the plant to the support.

Wire netting cylinders or shapes, supported with wooden stakes or canes, make excellent frames to support Ivies which eventually completely cover these supports.

Composts

The majority of the plants listed here, grow best in a peat based compost, i.e. Levingtons or Westland. The multi-purpose type offers the best of both worlds, being light, clean, easy to handle and is ideal for both mature specimens, young cuttings and seedlings. Camellias, Leptospermums and Azaleas are best grown in lime-free compost (ericaceous). All potting composts contain 6-8 weeks of food but thereafter will require liquid feeding. A good all round food ideal for flowering and foliage plants is 'Miracle-Gro' or 'Miracid' for lime hating plants.

LOVELY AND GOOD LOOKING HOUSEPLANTS

ALOE

Aloe (*Aloe* spp.), an easy-care succulent, has distinctive elongated leaves that fan out in a vase shape from a central base. Try smaller varieties such as *Aloe vera*on a sunny kitchen window. Aloes work nicely in dish gardens and in rooms with Southwestern decor. Keep the spiky leaves away from high-traffic areas. Size: To 3 feet high and 3 feet wide Growing conditions: Bright light; 65 - 75°F; moderately dry soil

Anthurium

Showy flowers give bright color for eight weeks or more each year to anthurium (*Anthurium andraeanum*, also known as flamingo flower). Flowers typically are red, but you can find hybrids in shades of pink, lavender, white and even green.

Anthurium flower blossoms make superb cut flowers because they last a long time. Note that the pretty, heart-shape leaves contain toxic sap, so make sure pets and children do not ingest them. Size: To 2 - 3 feet high and 2 feet wide Growing conditions: Medium to bright light with no direct sun; 65 - 80°F; evenly moist soil (barely moist in fall and winter)

Pictured: Anthurium *(Anthurium andraeanum)*

Asparagus fern

Delicate needlelike foliage drapes from this subtropical relative of asparagus, which is not a true fern. Stems of asparagus fern *(Asparagus densiflorus)*shoot upward and outward, making it a good hanging basket plant. Small white flowers hidden among needles turn into bright red berries. Keep the poisonous berries away from children and pets.

Size: To 18 - 36 inches high and 12 - 36 inches wide. Growing conditions: Medium to bright light; 60 - 75°F; evenly moist soil

Pictured: Asparagus fern (*Asparagus densiflorus* 'Sprenger')

Peace lily

The easy-care peace lily *(Spathiphyllum wallisii)* tolerates low humidity and low light. Its glossy, lance-shape leaves tip arching stems that surround the central flower spikes. The spoon-shape flowers normally appear in summer, but many cultivars bloom intermittently throughout the year. The dark leaves look attractive in a plain pot with a glossy finish.

Size: To 1 - 6 feet high and 1 - 5 feet wide Growing Conditions: Low to bright light; 60 - 85°F; evenly moist soil

Pictured: Peace lily *(Spathiphyllum wallisii)*

Peperomia

Peperomia's most interesting feature is its leaves, which vary in shape (heart shape to narrow), texture (waxy to wafflelike) and color (green, reddish or silvery gray). Plants occasionally produce slender flowery spikes that resemble rat's tails. Use on tabletops and as a companion in dish gardens and mixed baskets. The plant tolerates the low light of a north windowsill and stays small enough to fit on a desk or to be used in a terrarium.

Size: To 6 - 12 inches high and 6 - 12 inches wide

Growing conditions: Low to medium light; 60 - 75°F; moderately dry soil

Pictured: Watermelon peperomia*(Peperomia argyreia)*

Snake plant

This carefree, tough succulent grows almost anywhere. Snake plant (*Sansevieria* spp.) tolerates neglect but responds nicely to good care. Leathery, sword-shape leaves grow edged with yellow or white. Snake plant is great for beginners, but experienced houseplant growers also love it for its dramatic upright form. When grown in bright light, it sends up a tall stalk of greenish fragrant flowers. The dwarf rosette varieties make nice desktop or tabletop plants.

Size: To 6 - 48 inches high and 6 - 36 inches wide

Growing Conditions: Low to bright light; 60 - 85° moderately dry soil

Pictured: Variegated snake plant (*Sansevieria trifasciata* 'Laurenti')

Cast-iron plant

Slow-growing cast-iron plant *(Aspidistra elatior)* lives up to its name. It's almost indestructible, withstanding neglect, low light, low humidity and a wide range of temperatures — perfect for a dark corner. Start with large plants since plants grow slowly.

Size: To 1 - 2 feet high and 1 - 2 feet wide

Growing conditions: Low light; 45 - 85°F; evenly moist soil (barely moist in fall and winter)

Pictured: Cast-iron plant *(Aspidistra elatior)*

Christmas cactus

One of many types of easy-care cactus, Christmas cactus *(Schlumbergera x buckleyi)* offers a graceful arching appearance, with long segmented stems and whorls of satiny flowers in lilac, deep rose, salmon, red-orange or white. Plants usually bloom mid- to late December. After blooming is finished, prune by pinching or using a sharp knife to cut off several sections. This encourages the plant to branch, creating a fuller plant with more blossoms.

Size: To 8 - 12 inches high and 6 - 18 inches wide

Growing conditions: Bright light; 70 - 80°F (55°F in fall); moderately dry, well-drained soil

Pictured: Christmas cactus*(Schlumbergera x buckleyi)*

Dieffenbachia

Arching pointed leaves up to 12 inches long, usually marbled with white or cream, grow out of a canelike stem. The large leaves of dieffenbachia (*Dieffenbachia* spp.) provide a tropical architectural accent; the plant also blends well into a mixed grouping of foliage. One of dieffenbachia's common names, dumb cane, comes from the toxic sap in the leaves and stems that causes tongue numbness and swelling when chewed by humans or pets.

Size: To 1 - 6 feet high and 1 - 3 feet wide

Growing conditions: Low to medium light; 65 - 80°F; evenly moist soil

Pictured: Dieffenbachia *(Dieffenbachia seguine)*

Chinese evergreen

An excellent foliage plant for low to medium light, slow-growing Chinese evergreen *(Aglaonema commutatum)*can live for 10 years or more. Its standout feature: arching, lance-shape leaves, which are usually variegated with silver, gray or shades of green. Put a small plant on a living room table, or group larger ones with other low-light plants.

Size: To 1 - 3 feet high and 1 - 3 feet wide

Growing conditions: Low to medium light; 60 - 75°F; evenly moist soil

Pictured: Chinese evergreen*(Aglaonema commutatum)*

Dracaena

A large group of popular foliage plants, Dracaena (*Dracaena* spp.) display long, straplike leaves variegated with white, cream or red. Use young plants on tabletops. Larger plants require floor space, where they make striking specimens or work well in foliage groupings. Maintain plants at less than 6 feet by lopping off the top of the plant; within weeks a pair of new shoots will appear just below the cut.

Size: To 1 - 10 feet high and 1 - 3 feet wide

Growing conditions: Medium to bright light; 65 - 75°F; barely moist soil

Pictured: Corn plant (*Dracaena fragrans*'Massangeana')

English ivy

This versatile foliage plant's dainty demeanor makes it suitable for hanging baskets or pots. It's among the best houseplants for training on topiary forms or using as a groundcover beneath larger houseplants. Place English ivy *(Hedera helix)* on a mantel or shelf where the stems can hang down. The stems can grow quite long but are easily controlled with pruning.

Size: To 6 - 12 inches high and 6 - 72 inches wide, trailing

Growing conditions: Medium to bright light; 55 - 70°F; average to high humidity; evenly moist soil

Pictured: English ivy *(Hedera helix)*

Ficus

Glossy green leaves spread out from an upright woody plant that's available as a single-trunk tree or multistem shrub. Use durable ficus (*Ficus* spp.) as a tabletop or floor plant depending on its size. Place your ficus where it won't have to be moved all winter, because it doesn't like changes in environmental conditions.

Size: To 1 - 12 feet high and 1 - 10 feet wide

Growing conditions: Medium to bright light; 65 - 75°F; barely moist soil

Pictured: Braided trunk form of weeping fig *(Ficus benjamina)*

Hoya

A fast-growing, trailing plant, hoya (*Hoya*spp.) shows off both smooth, shiny foliage and pink or white flowers with red centers. The waxy, sweetly scented flowers give hoya its alternate common name, wax plant. Place in hanging baskets or train it to grow upward along a trellis. Wrap long vines around a form to make a wreath. Small-leaf types can be trained into topiaries. The easiest of the clan: *Hoya carnosa*.

Size: To 6 - 12 inches high and 48 inches wide, trailing

Growing conditions: Medium (foliage only) to bright light (for owers); 55 - 75° moderately dry soil

Pictured: Hindu rope plant (*Hoya carnosa* 'Crispa')

Parlor palm

The elegant, upright parlor palm*(Chamaedorea elegans)* has been a popular houseplant since Victorian times. Straplike green leaflets on feathery fronds grow 9 to 24 inches long. It's great for filling an empty corner and can get by on lower light than other palms. Young plants will do nicely in a terrarium while small.

Size: To 1 - 8 feet high and 1 - 3 feet wide

Growing conditions: Medium to low light; 55 - 75°F; evenly moist soil

Pictured: Parlor palm *(Chamaedorea elegans)*

Philodendron

This durable foliage plant (*Philodendron*spp.) has long been the backbone of indoor gardening. The popular heart-leaf philodendron has slender stems with leaves up to 4 inches long. Philodendron is easy to grow, easy to propagate (just root from stem tip cuttings) and adapts well to almost any indoor setting. Try small specimens in dish gardens and mixed baskets. The plant's low light tolerance makes it a good choice for bookshelves and for draping over the sides of a large piece of furniture. Note that leaves are poisonous to pets and people if eaten in large amount.

Size: To 1 - 10 feet high and 1 - 6 feet wide

Growing conditions: Low to bright light; 60 - 80°F; moderately dry to evenly moist soil

Pictured: Heart-leaf philodendron *(Philodendron hederaceum oxycardium)*

Spider plant

Spider plant *(Chlorophytum comosum)*looks striking in a hanging basket or on a pedestal, with its "babies" overflowing the pot. Most plants are the cultivar 'Vittatum', which has a white stripe down the center of each leaf, though all-green plants are also available.

Brown leaf tips, which are common with spider plants, are due to contaminated water, usually from fluoride, or to overfertilization, low humidity or dry soil conditions. Trim damaged leaves with scissors. Use rainwater or distilled water if your tap water is fluoridated.

Size: To 6 - 12 inches high and 6 - 24 inches wide

Growing conditions: Medium to bright light; 65 - 75°F; evenly moist soil

Pictured: Variegated spider plant (*Chlorophytum comosum* 'Vittatum')

Tradescantia

The trailing stems of tradescantias (*Tradescantia* spp.) change direction slightly at each node, giving the plant a zigzag look. Grow in mixed baskets, hanging baskets or as a groundcover under larger houseplants. Varietaged and purple-leaf types need bright light to maintain colors. Older plants become leggy and unkempt; replace them with new plants from rooted cuttings.

Size: To 4 - 12 inches high and 12 - 24 inches wide, trailing

Growing conditions: Medium to bright light; 55 - 75°F; moderately dry soil

Pictured: Wandering Jew *(Tradescantia zebrina)*

Plectranthus

Despite its name, Swedish ivy*(Plectranthus verticillatus)*, the most common plant in the plectranthus family, is native to South African and Australia - not Sweden - and is more closely related to mint than ivy. Its cascading stems studded with glossy rounded, scalloped-edge leaves make it primarily a foliage plant. But if often blooms in late spring or early summer with tiny white or pale mauve blooms on short spikes. Its tidy drapes make Swedish ivy a great choice for hanging baskets.

Size: To 6 - 18 inches high and 6 - 36 inches wide
Growing conditions: Medium light; 50 - 65°F; barely moist soil

Pictured: Swedish ivy *(Plectranthus verticillatus)*

Arrowhead vine

This lush foliage plant retains its variegation — dark green leaves marked with white along the veins — even in low light. Use small plants in mixed baskets and with other foliage plants in tabletop gardens. Train vining types to an upright form on a moss pole. Arrowhead vine*(Syngonium podophyllum)* does well in artificial light.

Size: To 6 - 36 inches high and 6 - 36 inches wide
Growing conditions: Low to medium light; 60 - 75°F; evenly moist soil

Pictured: Arrowhead vine *(Syngonium podophyllum)*

PLANTS FOR REMOVING INDOOR TOXINS

Common indoor plants may provide a valuable weapon in the fight against rising levels of indoor air pollution. NASA scientists are finding them to be surprisingly useful in absorbing potentially harmful gases and cleaning the air inside homes, indoor public spaces and office buildings. The indoor pollutants that affect health are formaldehyde, Volatile Organic Compounds (benzene and trichloroethylene or TCE), airborne biological pollutants, carbon monoxide and nitrogen oxides, pesticides and disinfectants (phenols), and radon. These pollutants contribute to 'sick building syndrome', which causes symptoms ranging from allergies, headaches and fatigue through to nervous-system disorders, cancer and death.

Through studies conducted by NASA, scientists have identified 50 houseplants that remove many of the pollutants and gases mentioned above. Dr. B. C. Wolverton rated these plants for removing chemical vapors, ease of growth, resistance to insect problems, and transpiration (the amount of water they expire into the air).

NASA, with assistance from the Associated Landscape Contractors of America, conducted a two-year study directed by Dr. B.C. Wolverton, an environmental engineer from Picayune, Mr. Wolverton has worked as a research scientist for NASA for some 20 years. His study, in the late '80s and early '90s, of the interaction of plants and air found that houseplants, when placed in sealed chambers in the presence of specific chemicals, removed those chemicals from the chambers.

Areca Palm (Chrysalidocarpus lutescens)

Also called the "Butterfly Palm". An upright houseplant that is somewhat vase shaped. Specimen plants can reach 10 to 12 foot in height. Prefers a humid area to avoid tip damage. Requires pruning. When selecting an Areca palm look for plants with larger caliber trunks at the base of the plant. Plants that have pencil thin stems tend to topple over and are quite difficult to maintain.

Lady Palm (Rhapis excelsa)

Also called the "Lady Palm", this durable palm species adapts well to most interiors.

The Rhapis are some of the easiest palms to grow, but each species has its own particular environment and culture requirements. The "Lady Palm" grows slowly, but can grow to more than 142 in height with broad clumps often having a diameter as wide as their height.

Bamboo palm (Chamaedorea seifrizii)

Also called the "reed palm", this palm prefers bright indirect light. New plants will lose of some interior foliage as they acclimate to indoor settings. This plant likes to stay uniformly moist, but does not like to be over-watered or to sit in standing water. Indoor palms may attract spider mites which can be controlled by spraying with a soapy solution.

Rubber Plant (Ficus robusta)

Grows very well indoors, preferring semi-sun lighting. Avoid direct sunlight, especially in summer. Young plants may need to be supported by a stake. The Ficus grows to 8' with a spread of 5'. Wear gloves when pruning, as the milky sap may irritate the skin. Water thoroughly when in active growth, then allow the soil to become fairly dry before watering again. In winter keep slightly moist.

Dracaena "Janet Craig" (Dracaena deremensis)

The Dracaena grows to 10' with a spread of 3'. Easy to grow, these plants do best in bright indirect sunlight coming from the east/west. They can adapt to lower light levels if the watering is reduced. Keep the soil evenly moist and mist frequently with warm water. Remove any dead leaves. Leaf tips will go brown if the plant is under watered but this browning may be trimmed.

Philodendron (Philodendron sp.)

One of the most durable of all house plants. Philodendrons prefer medium intensity light but will tolerate low light. Direct sun will burn the leaves and stunt plant growth. This plant is available in climbing and non-climbing varieties. When grown indoors, they need to be misted regularly and the leaves kept free of dust. Soil should be evenly moist, but allowed to dry between watering.

Dwarf Date Palm (Phoenix roebelenii)

A hardy, drought-tolerant and long-lived plant, the Dwarf Date Palm needs a bright spot which is free of drafts. It grows slowly, reaching heights of 8-10'. The Dwarf Date Palm should not be placed near children's play areas because it has sharp needle-like spines arranged near the base of the leaf stem. These can easily penetrate skin and even protective clothing.

Ficus Alii (Ficus macleilandii "Alii")

The Ficus Alii grows easily indoors, and resists insects. It prefers a humid environment and low to medium light when grown indoors. The Ficus Aliii should not be placed near heating or air conditioning vents, or near drafts because this could cause leaf loss. Soil should be kept moist but allowed to dry between watering.

Boston Fern (Nephrolepis exaltata "Bostoniensis")

The Boston fern grows to 4' in height with a spread up to 5'. It has feathery ferns which are best displayed as a hanging plant. It prefers bright indirect sunlight. Keep the soil barely moist and mist frequently with warm water. This plant is prone to spider mites and whitefly which can be controlled using a soapy water spray. Inspect new plants for bugs before bringing them home.

Peace Lily (Spathiphyllum "Mauna Loa")

The Peace Lily is a compact plant which grows to a height of 3' with a 2' spread. This hardy plant tolerates neglect. It prefers indirect sunlight and high humidity, but needs to be placed out of drafts. For best results, the Peace Lily should be thoroughly watered, then allowed to go moderately dry between waterings. The leaves should be misted frequently with warm water.

SPIDER PLANTS AND CLEAN AIR

Perhaps you remember the news from more than a decade ago that spider plants seemed to do a spectacular job of cleaning the air. The National Aeronautic and Space Administration (NASA), which tested the abilities of three common houseplants to remove formaldehyde from the air, found in preliminary tests that spider plants were the champs, removing 95 percent of the toxic substance from a sealed Plexiglas chamber in 24 hours.

Since then, much more research on the phenomenon has been conducted, and the news about spider plants is no longer so rosy. But that´s not because the tests were wrong. It turns out other plants also clean the air, of contaminants ranging from ammonia to benzene. Such toxics waft from countless sources in our daily lives, including adhesives, ceiling tiles, photocopiers and upholstery. And some plants do a much better job than the spider plant. The growing body of research on the subject offers yet more evidence that houseplants may be an effective means of improving the air quality in your home and office.

Much of the research has been conducted by former NASA research scientist Bill Wolverton, who was first interested in greenery as part of ecological life-support systems for space travel. Wolverton found that plants absorb airborne pollutants as part of their normal "breathing" process (carbon dioxide in, oxygen out) and transport them to their roots, where microbes feed on and detoxify them.

Wolverton has measured the filtering capacity of specific plants by putting them into sealed, 85-gallon chambers with a succession of toxic chemicals, then measuring how much of each chemical remains at 12-hour intervals. The results have been impressive. "We found that the most effective plants had a high transpiration rate—that is, they emit more moisture into the environment," says Wolverton. "This is the result of a natural pumping action. As plants pull air down to their roots, they pump moisture out."

Now retired from NASA, Wolverton is promoting houseplants in his book *How to Grow Fresh Air* and through his business, Wolverton Environmental Services in Picayune, Mississippi. Soon he will be marketing an "enhanced" planter with a fan and filter designed to maximize a plant´s natural pumping action. But while other studies have verified and extended his findings, the precise filtering capacity of plants outside the laboratory remains an open question.

"There´s no argument about what plants do," says John Girman of the Environmental Protection Agency´s Indoor Environments Division. "It´s a question of how well they do it, and whether that´s enough to be an effective means of pollution control in a typical indoor environment." While one plant may clean the air in a small, confined space, getting those same results for a whole house would require many more plants. Just how many is not clear. Wolverton figures 60 would do the job; Girman thinks the number is closer to

700. Part of the uncertainty comes from the question of ventilation. In sealed chambers, plants have filtered as much as one-tenth of the air every hour. "That sounds pretty good until you compare it to the ventilation in an average home," says Girman, who calculates that three-quarters of that air is replaced every hour. "In other words, that great result you´re getting in a sealed chamber isn´t anywhere near the normal air turnover you get in a typical home," he says. Wolverton, however, says the typical air change in a house is more like 40 percent an hour. And some structures are far more airtight. "Tightly sealed buildings create conditions very similar to static lab tests in that there is very little movement of air," he maintains.

Another concern is the static nature of lab tests, in which a plant processes a limited amount of toxic material. Most indoor air pollutants are continually emitted by their sources. "In real-life situations, those toxicants are constantly being replenished," says environmental scientist Thad Godish of Ball State University. "In fact, when you lower the concentration of toxicants in the atmosphere, the change in vapor pressure actually induces a source like particleboard to release them at a faster rate."

Wolverton counters that in tests he and colleagues have found that a plant can continuously remove formaldehyde from the air. Not only that, he has found that over time at least one plant, the lady palm, actually becomes more efficient at disposing of the gas.

The anecdotal evidence suggests benefits beyond anything demonstrated in the laboratory. "I get calls from people all the time who tell me their headaches or sore throats or other ´sick building´ symptoms have either gone away or greatly diminished after they´ve brought in plants," says Jan Roy, managing director of the Plants for Clean Air Council, an advocacy group funded in part by private growers.

Because each plant has an affinity for different toxics, Roy and Wolverton recommend using a mix of plants. "The more you have, the faster they´ll work," says Wolverton, who has recommended, as a rule of thumb, two or three plants in 8- or 10-inch pots for every 100 square feet of floor space.

One potential drawback of having many plants is excessive humidity, which can contribute to the growth of mold and bacteria and can encourage unwanted pests. To help prevent that, don´t let your plants sit in stagnant water. Once water has drained through the soil into the pan or tray below, dispose of it. Sub-irrigation planters, which have built-in watering tubes, bypass the problem altogether by funneling water directly to a reservoir in the soil. You can discourage mold from growing on top of the soil by covering it with Spanish moss or aquarium gravel.

Plants should never be thought of a cure-all, Wolverton cautions. "If you have a new building with serious indoor air pollution, even a large number of plants isn´t going to solve the problem," he says. But like a cat in your lap, a plant in your "personal breathing zone"—which Wolverton defines as the six to eight cubic feet of air surrounding an individual—can be a comforting presence.

"If you spend a lot of time at a computer or in an easy chair watching TV, try to have a nice plant sitting near you," he suggests. "Because the closer you are to that plant, the more it´s going to benefit you."

HOUSEPLANTS TO IMPROVE INDOOR AIR QUALITY

Fig. Golden Pothos (Epipiremnum aureum)

We all remember learning in science class that plants "breathe" by taking in carbon dioxide and releasing oxygen, which is why forests are so important in maintaining the delicate balance of the earth's atmosphere. But did you know that certain tropical houseplants can also remove and process other, more harmful, chemicals from the air inside your home?

Fig. Dracaena deremensis 'Warneckii'

Toxic Chemicals in Household Air

Homes and office buildings today are often more polluted than the outdoors. Modern buildings are tightly sealed and insulated to make them energy efficient, and they're also full of synthetic materials that emit harmful gases and chemicals into the air. The result is a sealed bubble of unclean air that can lead to what is known as "Sick Building Syndrome." Some of the most common indoor pollutants are:

- Formaldehyde: Commonly used in a number of items including particle board, pressed wood, foam insulation, cleaning products, and treated paper or fabric. If your home or office contains particle-board furniture, grocery bags, tissues, paper towels, or anything that has been treated to make it stiffer, wrinkle-resist, fire retardant, or water-repellent, then you're likely to have formaldehyde in the air.

- Benzene: A solvent used in manufacturing paints, inks, plastics, rubber, dyes, pharmaceuticals, and detergents.
- Trichloroethane: Can be found in adhesives, varnishes, paints, and used in dry-cleaning.

At the very least, chemicals like these can irritate the eyes and skin, lead to allergic reactions, and cause headaches. At worst, they've been linked to more serious problems including asthma, cancer, anemia, organ damage, and birth defects. Given the pervasive presence of these chemical in our homes, it can be difficult to create an environment that is free of them.

Fig. Ficus tree (Ficus benjamina)

Research by NASA and ALCA

In the late 1980s, a two-year research study was conducted by NASA and the Associated Landscape Contractors of America (ALCA) to investigate ways to create healthy, breathable environments in outer space. They found that certain tropical plants, commonly used as houseplants, were quite effective in removing formaldehyde, benzene, and trichloroethane from the air and replacing it with breathable oxygen.

All plants provide some benefit to air quality, but their research showed that tropical plants (grown as houseplants in cooler climates) are particularly effective at processing gases and chemicals. Because they grow in dense rainforests with very little light, they have evolved to be very efficient at photosynthesis, which includes the absorption of gases from the air. In addition, as plants transpire (emit water from the leaves), air is drawn down around the roots, where root microbes quickly adapt and begin "eating" the harmful chemicals that are absorbed.

The result was a list of recommended plants for reducing toxic chemicals in indoor environments. Most are common houseplants that you should be able to find at your local garden center. Two of the recommended plants (Gerbera

Daisy and Pot Mum) are ornamental blooming plants that are frequently brought indoors for seasonal decorations.

Fig. Heartleaf Philodendron (Philodendron scandens 'Oxycardium).

Top Plants for Reducing Indoor Air Pollution

- Bamboo Palm (Chamaedorea seifrizii)
- Chinese Evergreen (Aglaonema modestum)
- Dracaena (Dracaena sp.). Especially Red-Edged Dracaena (Dracaena marginata), Janet Craig (Dracaena deremensis 'Janet Craig'), Warneckii (Dracaena deremensis 'Warneckii'), and Corn Plant (Dracaena fragrans 'massangeana')
- English Ivy (Hedera helix)
- Ficus, or Weeping Fig (Ficus benjamina)
- Gerbera Daisy (Gerbera jamesonii)
- Golden Pothos (Epipiremnum aureum)
- Peace Lily (Spathiphyllum sp.)
- Philodendron (Philodendron sp), particularly Heartleaf (Philodendron scandens 'oxycardium'), Selloum (Philodendron selloum), and Elephant Ear Philodendron (Philodendron domesticum)
- Pot Mum (Chrysanthemum morifolium)
- Snake Plant, or Mother-In-Law's Tongue (Sansevieria trifasciata 'Laurentii')
- Spider Plant (Chlorophytum comosum)

For best results, have at least one six-inch plant for every 100 square feet.

Fig. Peace Lily (Spathiphyllum sp.)

Other Benefits of Indoor Plants

Of course a home or office building cannot exactly mimic the controlled conditions of a research lab, but it is clear that having plenty of healthy houseplants leads to a healthier home. In addition to processing carbon dioxide and harmful chemicals, houseplants improve indoor air quality by:

- Helping to maintain humidity levels: Dry indoor air is blamed for a host of respiratory problems, particularly in the winter, and plants emit water vapor during transpiration.
- Producing negative ions: Plant leaves produce negative ions, similar to many air purifying machines. Negative ions attach themselves to (and effectively remove) particles such as dust, mold spores, bacteria, and allergens. The presence of negative ions is credited for increasing psychological health, productivity, and overall well-being.

Fig. A variegated form of English Ivy (Hedera helix)

The Other Side of the Debate

Some researchers, including the EPA Indoor Air Division, remain unconvinced that houseplants are the answer to cleaning indoor air. The reasons for their doubt include:

- Houseplants should not be your only defense against indoor air pollution. Instead, pollution should be eliminated at its source by reducing the amount of synthetic material in your home or office, and by making sure buildings are well ventilated.
- The average home or office building is different from a controlled research lab, so it's difficult to determine if the results translate to the real world, or how many plants are needed to get the same effect. While no one disagrees that plants process chemicals in the air, their exact rates and effectiveness are hard to prove outside the lab setting. Some critics believe it might take hundreds of houseplants to get the same results as the NASA/ALCA study.
- Too many plants can raise humidity levels too high, which could lead to mold and bacteria growth throughout the building. Indoor humidity levels should stay between 35%-65% to avoid turning your home into a "greenhouse."

- Moist soil breeds bacteria, mold, and mildew. Don't overwater your plants, and help control mold by "mulching" your houseplants with a one-inch layer of fine gravel or other porous material.

Fig. Corn Plant (Dracaena fragrans 'Massangeana')

CLEAN THE AIR IN YOUR HOME WITH HOUSE PLANTS

Science is now catching up with what gardeners have known for decades — that is, growing plants can relieve stresswhile helping to clean the environment. Gardening has become the number one leisure activity in the United Statesand Canada, surpassing even sports. A growing body of research shows that cultivating plants indoors and outdoorsmay be the best medicine available for improving mental and physical well-being at any age.

Although "green building" is becoming an attractive concept to building managers and building occupants, the use ofliving plants is not part of the present concept. Architects and engineers are beginning to design buildings with aneye toward low-emitting carpets, paints and furniture. This is good but should only be the first step. A further stepshould include the design of houseplants into each building, mimicking the earth's natural processes.

Benefits from our botanical friends

Benefits derived from our botanical friends include a wide range of psychological and physiological effects. Studiesconducted on plant/people interactions have provided overwhelming evidence that plants do indeed have ameasurable beneficial effect on people and the spaces they inhabit.

Plants not only add beauty to a room, but also make it a friendly, inviting place to live or work. Plants symbolizefriendship and appear to have a calming, spiritual effect on most people. This perhaps explains why plants play suchan important role in human events such as weddings, funerals, holidays, hospital stays and birthdays.

Plants are also used as background props for most important events such as television addresses, commercials, etc.People feel relaxed when they are near or tending to living plants. Corporations install interior landscaping to increaseworker productivity and decrease absenteeism. Elite hotels, restaurants and other businesses use plants to helpentice customers to their establishments.

During early manned space flights, NASA astronauts and Soviet cosmonauts expressed a desire to have plants onboard their space vehicles. Plants can help reduce stressful conditions inside cramped space capsules duringlong-duration flights.

NATURE'S BIO-CLEANING MACHINES

In the past, houseplants were sought only for their beauty and psychological value. Thanks to NASA researchfindings, houseplants now have a third value. Studies conducted in the early 1980s at the John C. Stennis SpaceCenter in Mississippi provided evidence that houseplants can also improve indoor air quality. The ability ofhouseplants to improve indoor air quality and one's health is no longer a matter of conjecture – it's scientific fact.

Plants and their root microbes are nature's biological cleaning machines. It is commonly understood that plants purifyand revitalize the earth's air and water. In general, we know that the animal/plant/microbial world is harmoniouslybalanced so that each benefits from the other. We are dependent upon these interactions for our existence.

We are just now beginning to understand some of the mechanisms that create these symbiotic relationships.Approximately 42 species of interior plants have been evaluated for their ability to remove various indoor aircontaminants from sealed chambers. Hundreds of experiments have been conducted and technical reports publishedthat seek to answer legitimate concerns about placing plants in buildings for the specific purpose of improving indoorair quality.

After more than ten years of extensive research (both laboratory and "real-world), we now have a basicunderstanding of how plants function to remove indoor pollutants. Research conducted by Wolverton EnvironmentalServices, Inc. and supported by the Plants for Clean Air Council in Mitchellville, Maryland, continues to expand on theresearch begun at NASA. Specifically, we are trying to understand how plants clean and revitalize the air and how touse this knowledge to improve indoor air quality.

Plants use ingenious methods to obtain food and protect themselves from would-be enemies. Each plant has theability to culture microbes on and around its roots specific for its needs. These microbes biodegrade and mineralize(compost) dead leaves, animal waste, tannic and humic acids and other debris to provide nutrients for the microbesand their host plant. This is basis of organic gardening.

Geographic locations and environmental conditions of the plant's origin determine which microbes it cultures. Forexample, the microbes associated with plants that evolved underneath the canopy of tropical rainforests (mosthouseplants) differ from those in arid environments. Tropical plants need aggressive microbes that can rapidly recyclejungle debris. Because rainforests are dark, warm and humid, mold and bacteria thrive.

Tropical plants excrete substances that protect their leaves from airborne molds and mildew. When these plantspecies are placed in an indoor environment, they continue to suppress airborne mold spores. Because chemicalpollutants commonly found indoors such as formaldehyde, benzene and xylene have structures similar to componentsfound in tannic and humic acids, microbes adapt to biodegrade these chemicals also. Thus, the basis for plants'ability to improve indoor air quality is established.

Humidity: the basics

Plants use two well known processes to move chemicals in the air to their roots: Leaves absorb certain chemicals inthe air and transport them inside plant tissue down to the roots, and plants pull air down around their roots whenmoisture is emitted from leaves during transpiration.

Plants with high transpiration rates are able to move greater amounts of air. Therefore, the more efficient air cleanersare plants with high transpiration rates. Plant transpiration rates are controlled by humidity. Plants attempt tobalance humidity levels for their optimum well-being by controlled release of moisture from their leaves. Whenhumidity is high, plants emit less moisture into the air then when humidity is low.

Early critics complained that too many plants in buildings would cause the humidity levels to rise and support thegrowth of mold and mildew. However, findings proved otherwise. Low humidity, most prevalent during winter months,dries the respiratory system and makes one more susceptible to colds, viruses and allergens. Ideally, humidity shouldrange between 40 to 60 percent. Plants produce healthy, microbial-free moisture.

Mechanical humidifiers, when not properly maintained, can become a source of mold and mildew.When plants transpire, they not only add moisture to the air but also emit substances that help suppress airbornemold spores and bacteria. Although these substances are yet to be identified, we do understand their function.Recent findings show that plant-filled rooms contained 50 to 60 percent fewer airborne mold and bacteria than roomswith no plants. Interestingly, air in the plant-filled rooms had fewer microbes, even when temperature and humiditylevels were raised — the exact opposite effect predicted by some critics.

Ironically, some doctors advise their allergy patients to avoid house plants. House plants have been falsely accusedof harboring mold spores. The real problem is usually overwatering and the growth of mold on wet carpeting. To avoidthese problems, use hydroponic (soil-less) methods in water-tight plants to grow house plants. If potting soil is used,cover it with aquarium gravel and feed and water from the bottom to keep the surface dry. There are also manycommercial sub-irrigation systems available. When large planters are used, the need for frequent watering can beeliminated.

Healthy air for your home

As a general guide, two or more medium to large plants (14"-16" containers) per 100 square feet of area arerecommended. Of course, more plants and larger plants would certainly increase effectiveness.

Plants alone may not be the total solution when serious indoor air quality problems exist. Proper source management(allowing building materials and furnishings to vent volatile fumes before installation), complete air distribution andpreventive maintenance are all components of a healthy building.

Fan-assisted planter/air filters

Fan-assisted planter/filters may be needed to rapidly remove pollutants from the air. Once the biological mechanismsof plants were understood, it was only natural to merge man and nature's technologies. By combining the mosteffective air filtering media, mechanical air flow devices and living plants, WES Inc. has developed a family ofenhanced plant/air filters. These aesthetically designed, patented planter units not only increase the air purificationcapacity of houseplants by as much as 200 times but also help maintain healthy indoor humidity levels.

One of the unique properties of this natural air purification is that under normal operating conditions, the filteringmedia is bioregenerated (self-cleaned) by the plant's root microbes. Therefore, the filter media does not requireperiodic replacement, as is the case with other commercial air filters.

We should all breathe easier knowing our beautiful house plants are working so hard to keep us healthy!

PLANTING HEALTHIER INDOOR AIR

Poor indoor air quality has been linked to health problems, especially in children. Asthma has reached epidemic proportions among multiple age groups and is considered the most common chronic disease in urban-dwelling children. The American Academy of Allergy, Asthma and Immunology Indoor Allergen Committee suggested in a 2010 report that allergists consider indoor air filtration to be part of a comprehensive strategy to improve respiratory health.2 Air cleaners with HEPA filters have been shown to improve symptoms of asthma. However, filtration systems and air purifiers do not reduce levels of all indoor air pollutants, and some types can actually aggravate the problem. For example, one study showed that some air purifiers raise indoor concentrations of ozone above safety levels established by the U.S. Environmental Protection Agency.

A more benign addition to air filtration could be the use of houseplants. In addition to basic photosynthesis that removes carbon dioxide and returns oxygen to the air, plants can remove toxicants from air, soil, and water in at least two ways. First, they can metabolize some toxic chemicals, releasing harmless by-products, and second, they can incorporate toxicants such as heavy metals into plant tissues, thus sequestering them.

Data on plant-mediated indoor air quality come from experiments conducted by the U.S. National Aeronautics and Space Administration (NASA). As NASA researchers explored the possibilities of long-term space habitation, it became evident that the air in a tightly sealed space capsule would quickly become contaminated with volatile organic compounds (VOCs) and other chemicals released by the materials used to manufacture the capsule interior.

This is similar to the situation in newly constructed energy-efficient dwellings. If energy-efficient construction is not carefully designed to maintain indoor–outdoor air exchange, one unintended consequence can be increased concentrations of pollutants indoors. For example, in a study recently published in the*American Journal of Public Health*, Gary Adamkiewicz and colleagues used a simulation model to demonstrate that in homes with low air exchange rates and multiple sources of air pollution, up to 90% of exposure to fine particulate matter came from indoor sources. Besides particles and VOCs, indoor air and dust can also contain brominated flame retardants, pesticides, toxic metals, and other pollutants.

For more than 30 years, B.C. "Bill" Wolverton, a retired civilian scientist for NASA, investigated the use of plants as air- and water-purifying systems for enclosed environments in space missions. Through his research, Wolverton found the air-cleaning capacity of houseplants can be improved exponentially by increasing air circulation to the roots of the plants, where symbiotic microorganisms help make the substances culled from air bioavailable to the plant.

For maximum benefit, multiple species of houseplants would likely be needed on a site to remove the relevant toxicants in a particular space, given that houseplants vary in the types of chemicals they are able to remove from the environment and the efficiency...

In those studies, Wolverton and colleagues tested several types of low-light houseplants. For example, golden pothos (*Epipremnum aureum*, also known as devil's ivy) grown on an activated carbon filter system reduced air levels of benzene and trichloroethylene inside a Plexiglas chamber measuring 0.58 cubic yard from approximately 36 ppm to barely detectable levels within 2 hours. Experiments conducted elsewhere by Stanley J. Kays and colleagues at the University of Georgia also documented the ability of different plant species to remove VOCs such as benzene, toluene, octane, and trichloroethylene.

One indoor contaminant of particular concern is formaldehyde, which is released by many household products, among them pressed woods, some types of foam insulation, paper products, some paints and varnishes, and permanent-press fabrics. The National Toxicology Program lists formaldehyde as reasonably anticipated to be a human carcinogen.

In an unpublished 2006 study, Wolverton tested a small fan-assisted planter/air filter inside a travel trailer that had been used as temporary housing for displaced Hurricane Katrina victims. This trailer, like similar units, had been found to be highly contaminated with formaldehyde. The plant/air filter contained a plant growing in a mixture of activated carbon and expanded clay pebbles. Wolverton's tests showed that the levels of formaldehyde were reduced from potentially toxic levels of 0.18 ppm to 0.03 ppm, within the safety limits defined by the World Health Organization.

Those studies fit well with evidence on the biochemical mechanisms involved in plant detoxification of formaldehyde. In studies published this year Zhongjun Xu and colleagues tested three kinds of potted plants for their capacity to remove formaldehyde from indoor air in test chambers. They found that the formaldehyde-removal capacity of the plants depended on the dehydrogenase activity in the leaves and root system—that is, how efficiently the plant could metabolize formaldehyde. As Wolverton found earlier, these investigators also found that formaldehyde removal by plants was diffusion-limited. That means increasing the circulation of contaminated air through the root system and leaves improved the formaldehyde-removal effect.

In another recent study, Kays and colleagues tested 86 species of houseplants from five general classes for their ability to remove formaldehyde. In their experiments, ferns had the highest formaldehyde-removal efficiency of all the plants tested, especially *Osmunda japonica*, commonly known as Japanese royal fern, or zenmai.

Another important air contaminant that is amenable to plants' cleanup abilities is mercury vapor. Mercury can make its way into homes through accidental spills (for instance, breakage of thermometers and fluorescent bulbs) as well as through its use in certain cultural and religious practices. Mercury vapor is neurotoxic and lingers in the air even after new sources have been eliminated from the environment.

Joao Paulo Machado Torres, a senior scientist at the Radioisotopes Laboratory of the Federal University of Rio de Janeiro, Brazil, and his group have published many studies on the use of plants in indoor and outdoor mercury-contaminated settings. "We have used plants of the bromeliad family and Spanish moss (*Tillandsia usneoides*) as sentinel species to detect and absorb mercury from the air in shops contaminated by the gold trade in the Amazon," he says. The use of plants can be uniquely useful in these environments where other kinds of remediation technology may be impractical or difficult to deploy.

But as has been shown with many natural remedies, "natural" does not necessarily equate to "absolutely harmless." A study by Kays and colleagues published in 2009 pointed out that some houseplants—as well as the media and plastic pots they are grown in, the microorganisms that inhabit them, and the pesticides used to treat them—can potentially contaminate indoor air with

VOCs. "It is not yet possible to project the true potential of plants for purifying indoor air," Kays says. "At this time the role of plants, though appearing [generally] positive, is not totally clear. The absence of funding for phytoremediation research has greatly impeded solving the problem."

Kays also notes the lack of an accurate means for the public to determine if the VOCs in their home or office represent a significant health problem. "The absence of a relatively inexpensive method available to the public results in situations where it takes two and a half years to determine that the Katrina trailers had toxic levels of formaldehyde even though there had been health complaints by the occupants almost as soon as the trailers were in place," he says. "If an accurate, reasonably priced method was available from a credible source such as a university extension analytical laboratory, the public would be able to ascertain their potential health risk before buying or renting a house, apartment, or office."

5

Indoor Plants for Foliage and Flowers

FOLIAGE PLANTS

WATERING

This cultural practice generally causes the most confusion, but is relatively simple. Foliage plants are adapted to regions where soil is moist, but not continually saturated with water. Therefore, never allow soils to become completely dry between waterings, and when waterlog, apply enough water to thoroughly wet the entire soil ball. One of the best ways to do this in large plantings is with automatic watering systems. Watering frequency will depend on weather, type and size of plant and stage of plant growth.

The type of container used will affect method of watering. Containers with drainage holes should be placed in a saucer or other container and sufficient water applied at each waterlog until excess water drains out the bottom. This water should then be discarded, since pots should not be allowed to stand in water for any length of t1me.

Containers without drainage should have a layer of coarse gravel placed in the bottom to allow a space for excess water. A good way to check these containers for excess water is to lay them on their side in a sink or to use a dip-stick Inside a hollow pipe.

Soils and Repotting

Foliage plants can be grown in a great variety of soil mixes, but are easier to care for if planted in soils containing high levels of peat moss. Such so1ls retain water and fertilizer and provide good aeration if peat moss is mixed with coarse sandy soils or perlite. The following soil combinations are suggested for growing foliage plants, or prepackaged soils may be obtained with similar properties.

1. 2 parts peat, 1 part perlite, 1 part coarse sand.
2. 1 part peat. 1 part vermiculite, 1 part coarse sand.
3. 2 parts peat, 1 part coarse sand.

Repotting is necessary when the plant top outgrows the pot and there is not enough room for new root growth. Plants also require repotting when they must be watered more than once a day. When repotting, remove the loose soil. Use a clean pot of the right size. When shifting to a larger pot, place some soil in the bottom of the pot and firm the soil around the old root ball. Leave enough room in the top of the pot for proper watering. Water the soil thoroughly at first. Do not water the plant again until the soil dries out on top.

Fertilization

When foliage plants are used for interior design purposes, little fertilizat1on is required, since it is desirable to keep plants from growing excessively. Frequently, a lot of new growth is undesirable, since plants may soon out-grow their location or new growth may be unattractive if grown under less than optimum light conditions.

Application of fertilizers more than once a month is unnecessary. and in most cases four applications a year will be sufficient unless considerable new growth is desired. Many types of fertilizers are available for specialized Indoor use and are frequently more convenient than types sold for yards or gardens. Follow directions given on containers when using pre-packaged products sold specifically for "foliage plants". Many types of specialty fertilizers are available including liquids, tablets, and powders. Regular garden fertilzers, such as 6-6-6 or 8-8-8 can also be used; 1/8 to 1/4 teaspoon of either of these materials is sufficient for a plant in a 5-inch pot when applied every other month. More house plants suffer from an excess of fertilizer than from a lack of it. Over-fertilization will cause a slowdown of growth, stunted plants, burned or dried leaf margins and wilted plants.

Cleaning Foliage

When foliage plants are grown indoors, dust and other undesirable residues will accumulate on foliage and become unsightly, but most foreign matter can be removed by syringing or wiping foliage with a soft, moist cloth. A small amount of hand soap may be added to the water when stubborn spots are encountered. Don't fold, crease. or rub leaves too hard, as they may be damaged. Many people prefer plants with shiny foliage, a character than can be obtained by using a commercial "shine" compound. These are available in aerosol containers or as a liquid which is wiped on foliage. A cloth moistened with a little milk can also be used to obtain shiny foliage if rubbed gently over foliage of smooth leaved plants.

FOLIAGE PLANTS FOR YOUR HOME

Foliage plants that you grow indoors are mostly from tropical or arid regions and have to adapt to less than ideal conditions in your home or office. Your challenge is to know the plant's environmental needs and meet them. The environmental factors placed upon the plant and your maintenance practices will contribute to the health or decline of the plant, depending on how well you do.

Choosing Foliage Plants for Your Home

In choosing your plants, consider the location's environment. Determine whether a particular plant will only survive or thrive in that environment. Go to your library and find references of the specific care of plants being considered for the indoor environment.

It is important to begin with good quality, healthy, pest-free plants. Make sure leaves possess good color for the species, with no brown tips or margins.

ENVIRONMENTAL CONDITIONS FOR FOLIAGE PLANTS INDOORS

Lighting

How much or how little light in the environment often determines whether the plant will actively grow or simply survive. Characteristics of light to consider include intensity, quality and duration. Keep in mind that a southern exposure indoors typically provides the greatest light intensity, then western, eastern, and northern.

Plants requiring more light usually have variegated foliage. This is because they have less chlorophyll and therefore, require more light to achieve the same photosynthesis as a plant with green foliage. If light is insufficient, colorvariegation may be lost. Flowering plants also require higher light intensity.

As winter approaches, light intensity and duration will diminish. A plant that grew well in an eastern exposure in the summer may require a southern exposure in the winter. Move plants to other locations seasonally if needed. Quality of light refers to the spectrum or colors available; sunlight contains all colors. Plants utilize all colors in photosynthesis. An incandescent light bulb gives off limited colors and is not acceptable as an indoor lighting source for most plants. To grow plants under artificial fluorescent light, most indoor gardeners combine a cool and warm tube in a fixture to provide light of good quality for many interior plants.

Duration refers to the length of light exposure. A daily exposure to light, preferably eight to 16 hours, is needed for plant processes. Symptoms of insufficient duration are similar to those of low light intensity: small leaves, spindly stems and older leaf drop.

Temperature

The best temperature range for most interior foliage plants is between 60 and 80 F. (16-27 C.) These temperatures are similar to that found in the understory of a tropical forest. Chilling injury occurs below 50 F. (10 C.) for most tropical plants.

Temperatures in the home and office can be quite variable, changing daily or seasonally. Remember that southern and western exposures are warm because of sunlight, while eastern and northern are moderate or cool. Avoid locating plants on cold window sills, or where there are cold or hot drafts from opening doors and heating or air conditioning vents.

Leaf spots, blotches, downward curled foliage and slowed growth are all signs of bad temperatures. Temperatures that are too high can cause yellowish green foliage, which may have brown, dry edges or tips and spindly growth. Insect, mite and disease problems may develop quickly under warm conditions as well. You need to be careful.

Humidity

Remember that tropical foliage plants thrive in their native environments where relative humidity is often 80 percent or greater. An average home may have relative humidity as low as 35 percent up to about 60 percent; this may drop below 20 perccent in heated homes during winter. Low humidity may cause brown or scorched leaf tips. You can try to raise the humidity indoors by grouping plants together. That sometimes helps. Also, if you use a room or furnace humidifier, you can increase the humidity. Be sure to water properly and avoid drafts and high temperatures. A pebble tray may also work; layer pebbles in a tray and fill with water to just the top of the pebbles. Set pots on the pebbles, just above water level.

Soil

Root health is vital to the survival of the plant. The plant's container and the growing mix affect the root system and the overall health of the plant. Roots serve to anchor the plant in the container and to absorb water and nutrients. A plant's root system has to have oxygen in order to function properly. Without it, the plant will die. You have to make sure to have the proper soil mix for each plant as well. A good mix will not break down or degrade over time. You want to be sure to use a mixture of particle sizes so there is good drainage and aeration for the plant's roots. Most plants do well in a mix containing one to two parts potting soil, one to two parts moistened peat moss and one part coarse sand. Native soil from the garden can be used in a mix if it's pasteurized.

It's not hard to take care of foliage plants. You just have to remember that if they are tropical in variety, it might take a little more than just a simple watering once in a while to carry them through.

FERTILIZING FOLIAGE & FLOWERING PLANTS

MEETING PLANT NEEDS:

Plants brought into the store or home from the greenhouse are moving to an environment where light is less intense and hence where growth will be slower. Most plants already have sufficient fertilizer in their soil to sustain growth for up to three months under interior conditions. This is particularly true during fall and winter.

Even if plants continue to grow vigorously, it is wise to wait a month while it adjusts to its new environment, before adding fertilizer. Likewise, a newly potted plant needs time in which to grow new roots to absorb the fertilizer you will be applying.

It is often easier to feed a number of plants at the same time, and the schedules can be combined in this way: for a general practice, apply fertilizer to most indoor plants every 1-2 months while growth is rapid, and once or twice only during the winter months. Alternatively, apply the plant food more frequently in a dilute form; this accommodates plants which would be injured by full strength fertilizer. Approximately half the recommended strength means using only half as much fertilizer per plant or in solution.

Annual plants grow very quickly and need fertilizing every two or three weeks during their short season. Some flowering plants, like Azalea, should not be fed at all while they are in flower.

Fertilizer Types:

Any prepared mixture for the type of plants you have is fine: for most foliage and flowering plants, an N-P-K analysis ratio that is close to 1-2-1 will give balanced growth. African Violets grow best with a mix containing more nitrogen (N); and there are other formulations available for acid-loving plants like Azalea and Gardenia. For annuals and other garden plants growing in hanging baskets, an outdoor fertilizer is fine.

Soluble compounds are easy to apply and provide quicker results than solids or slow release pellets. Simply apply the fertilizer solution in place of water when you would normally be watering the plants. However, be sure the soil is slightly damp when you fertilize; never feed plants with dry soil, because roots will be damaged by the chemicals you add unless they are diluted further by soil moisture. Use all fertilizers at or weaker than the recommended strength of dilution, never stronger. Newly rooted cuttings, seedlings, and young plants will benefit from weaker solutions to avoid damage to the soft young roots.

Table. Fertilizers for foliage and flowering plants.

Name & Formula	N-P-K	Analysis Others
Ammonium Chloride NH_4Cl	25-0-0	0
Ammonium Nitrate $NH4NO_3$	33.5-0-0	0
Ammonium Phosphate (di) $(NH_4)2HPO_4$	21-53-0	0
Ammonium Phosphate (mono) $NH_4H_2PO_4$	11-48-0	1.4% Ca, 2.6% S
Ammonium Sulfate $(NH_4)2SO_4$	20-0-0	24% S
Calcium Nitrate $Ca(NO_3)_2$	15-0-0	37% S
Sodium Nitrate $NaNO_3$	16-0-0	0
Urea $CO(NH_2)_2$	45-0-0	0
Urea-Formaldehyde	38-0-0	0
Superphosphate $CaH_4(PO_4)_2$	0-20-0	18% Ca
Treble Phosphate $CaH_4(PO_4)_2$	0-42-0	12% Ca
Phosphoric Acid H_3PO_4	0-52-0	12% Ca
Potassium Chloride KCl	0-0-62	0
Potassium Nitrate KNO_3	13-0-44	0

GOOD INDOOR PLANTS WITH FLOWERS FOR LOW LIGHT

If you live in an apartment or you work in an office, sometimes you want to add color and greenery to the space. Low light indoor plants are an excellent way to add greenery, but what about color? A low light indoor plant with flowers is hard to come by but not impossible.

GOOD FLOWERING INDOOR PLANTS FOR LOW LIGHT

Consider some of these choices for flowers in low light:

Peace Lily – These low light indoor plants are more commonly seen in offices, but they make a great plant for your apartment as well. This indoor plant with flowers requires very little attention from you and needs very little light to flower. Peace lily will bloom year round with white-cupped, hand-shaped flowers. The flowers have a light scent and are long lasting. Advertisement

African Violet – Once thought of as a "grandma" plant,African violets have made a serious comeback thanks to a wide variety of new cultivars. These pretty plants make good indoor plants. They actually bloom best under neglect and need only minimal amounts of light in order to bloom.

Bromeliads – While technically this indoor plant is not a flower, the whole plant looks like a flower. The leaves of these low light indoor plants are vibrant and colorful and will add some flair to any room or cubicle. Bromeliads may also get real flowers, but in between, you can just enjoy their natural beauty.

Christmas Cactus – Christmas cacti make good indoor plants. Being a cactus also means that they need very little fuss to be happy. While they are called Christmas cactus, they don't actually always bloom at Christmas. Christmas cactus needs 12 hours of darkness to bloom, and this normally occurs during the winter months in most households. This is also why these plants are good low light indoor plants. The flowers on a Christmas cactus can range from white to pink to red.

Zebra Plant – Zebra Plants are other low light indoor plants that flower. The striped foliage looks nice year round and they will produce frilly, yellow flowers.

MOST POPULARFLOWERING AND FOLIAGE HOUSE PLANTS

Growing Requirements of the African Violet

African violets need about fourteen hours of *bright, indirect* light each day for the best flowering. It may be necessary to provide supplemental light for them by using a Gro-Light, especially during the winter months.

They should be planted in the smallest possible container because they need to be somewhat root bound to bloom. Grow them in a rich soil mix containing 1 part potting soil, 1 part peat moss and 1 part perlite or coarse sand. If you prefer, there are also many specialty mix African Violet soils also available. and certain trace elements that aren't available with an all purpose food. The A V food will take care of these specific needs.

The chlorine used in public water systems can be deadly to many houseplants. The chlorine will have dissipated by then, making it somewhat more safe for your plants. The water will also be at room temperature, which is much less of a shock to the plant. When watering African Violets, take care to keep water off of the foliage, flowers and crown of the plant.

Bottom watering is the best method, but never leave your plants sitting in water for more than an hour or so. This can quickly cause crown rot which is fatal. African Violets like a great deal of humidity *(40-60%)* but not wet leaves and stems, so misting is not an option. Setting the pots onto water splashed, pebble filled trays will often solve this problem, as will having an aquarium in the near proximity.

Spent flowers should be removed as soon as they begin to fade. This will allow the plants energy to be used to produce new blossoms rather than seeds.\ If you have a chance, try growing violets under fluorescent Gro-Lights. The colors of both the flowers and foliage will become more intense and bright.

Repot your African Violet plants annually with fresh soil. It is best to only increase the pot size by an inch or so. When you are working with African Violets use care not to touch the stems more than necessary, because they are by far more succeptable to damage than either the leaves or roots.

Propagating African Violets

African Violets are easy to propagate by division or by using leaf cuttings.

Leaf cuttings can be struck during any season of the year, but the winter months when the plant is less inclined to bloom seemed to work the best for me. Prepare a cutting tray or pot by filling it with a loose mixture of peat moss and vermiculite. Allow about 4 square inches of surface area for each cutting you intend to take.

Water the soil well to be sure that it is settled and moist. Using a nail or small stick, make evenly spaced, slightly angled cutting holes (the same depth as the length of your stems) Select a good healthy leaf and remove it (with stem) from the plant by reaching in as close to the crown of the plant as possible, grasping the stem and gently twisting it until it pulls free. With a sharp, sterile knife or razor blade, cut the stem at an slight angle *(matching the angle of the leaf)*, 1-1 ½" from the leaf.

Insert your cutting into your rooting media deep enough that about ¼-½" of the base of the leaf is covered by the soil mix. Gently press the soil around the cutting.Keep the rooting media evenly moist, but never soggy.

Place the tray in a warm (70°-75°), brightly lit area (not full sun). The cuttings should begin to produce roots in about two weeks, you should have a new plantlet in about eight weeks! Bottom heat and a grow light will speed up the process considerably.

Growing African Violets from Seed

Growing African Violets from seed is not as fast as growing them from leaf cuttings but it is a great way to add to your A V collection. Seed packets will usually contain many different varieties and colors of flowers.

Do not cover the tiny African Violet seeds with soil because they require light for germination. Sow the seeds on a bed of well draining, finely screened growing medium, then cover the tray with a pane of glass.

Maintain a temperature of 70°-75° within the growing medium. Germination takes 20-25 days but can be sped up with the use of fluorescent lighting.

Blooming takes up to eight months.

GROW AND CARE FOR ALOE VERA PLANTS

Fig. Aloe Vera (Aloe barbadensis)

There are over 250 species of Aloes in the world, mostly native to Africa. They range in size from little one inch miniatures to massive plant colonies consisting of hundreds of 2 foot diameter plants.

Although most Aloes have some medicinal or commercial value, the most commonly known is the Aloe barbadensis... better known as the Aloe Vera. All Aloes are semi tropical succulent plants, and may only be grown outdoors in areas where there is no chance of freezing (USDA zones 10-11). However, they make excellent house plants when they are given sufficient light. Container grown Aloe plants benefit from spending their summer outdoors. Older specimens may even bloom, producing a tall stock covered with bright colored coral flowers.

Growing Requirements for Aloe Vera Plants

Because Aloe Vera plants are very succulent and consist of 95% water, they are extremely frost tender. If they are grown outdoors in warm climates, they should be planted in full sun, or light shade. The soil should be moderately fertile, and fast draining. Established plants will survive a drought quite well, but for the benefit of the plant, water should be provided.

Because of their popularity, Aloe vera plants are available at almost every garden shop or nursery. Unless you live in area with a *very mild climate*, it's best to leave your Aloe plant in the pot and place it near a window that gets a lot of sun.

You can move the pot outdoors during the summer months. Aloe Vera is a succulent, and as such, stores a large quantity of water within its leaves and root system. During the winter months, the plant will become somewhat dormant, and utilize very little moisture. During this period watering should be minimal. Allow the soil to become completely dry before giving the plant a cup or two of water.

During the summer months, the soil should be completely soaked, but then be allowed to dry again before re-watering. Aloes have a shallow, spreading root system so when it is time to repot choose a *wide* planter, rather than a deep one. Use a planter with a drainage hole, or provide a 1-2 inch layer of gravel in the bottom of the pot to ensure adequate drainage. Use a good commercial potting mix with extra perlite, granite grit, or coarse sand added. You may also use a packaged 'cacti mix' soil.

Aloe Vera plants are propagated by removing the offsets which are produced around the base of mature plants, when they are a couple inches tall *(or larger)*. They can also be grown from seed.

The Medicinal Properties of Aloe Vera

The medicinal properties of Aloe vera have been known and recorded since biblical times. It has been used for a variety of ailments, and as an ointment for burns, cuts, and rashes, as well as an ingredient in various beauty preparations.

The sap of the Aloe is a thick, mucilaginous gel. It is this gel which is used medicinally. The outer skin has essentially no value, but because it is commercially easier and less expensive to utilize the entire leaf, 'whole leaf' Aloe juice has been hyped as the 'best'. This is not the case.

Using Aloe Vera Gel

As to the claims of the medicinal properties of the Aloe plant, I can only speak from my personal experience. I have kept an Aloe plant around for years, primarily for burns.

In case of burns, an immediate application of fresh gel has relieved much of the pain, and prevented blistering for me, many times. I also found it to be quite effective to relieve itching from stings, bites and various 'stinging' plants, such as poison ivy. Aloe gel is also good for the same problems when they are encountered by your pets.

When you need to use it medicinally, just remove a lower leaf from the plant, slice it open, and apply the gel on the affected area.

PILEAS AS HOUSE PLANTS

Pilea is a genus consisting of about 600 species of frost tender, tropical foliage plants that includes both upright bushy types and trailing varieties. Several smaller species are commonly cultivated as house plants because they are so easy to grow and care for.

These inexpensive little plants are excellent for growing in dish gardens and are great starter plants for young or inexperienced growers.Their foliage varies considerably, ranging from 3" strongly textured, lance shaped leaves to tiny heart shaped, moss-like foliage. Pileas occasionally bloom, but their pink or cream colored flowers are very tiny and often go unnoticed.

Growing Requirements for Pilea Plants

Pilea plants should be grown in bright, indirect light, but never in full sun. They will *thrive* under fluorescent 'grow lights'. Always keep your Pilea and any other house plants away from heating vents and out of drafts. Pilea plants grow best in warm daytime temperatures of 75°-80° with 10 degrees cooler at night.

Pileas need about 50% humidity in order to thrive, so they should be kept on a tray filled with moistened pebbles or misted regularly, or they can be grown in a terrarium. Keep the soil slightly moist while your Pilea is actively growing, but allow the top inch of soil to dry between waterings when growth slows in the winter. Feed monthly in the spring and summer using a half-strength solution

Pruning and Potting Pilea Plants

Pinch the tips of new growth on branching forms of Pilea to help keep the plant compact. Even with diligent pinching and regular clipping, Pileas tend to become leggy and unattractive because their lower leaves naturally drop with age. Luckily, Pilea cuttings root easily, so you can replace your old specimens with fresh new plants every year or two if you'd like.

Propagating Pileas with Stem Cuttings

Pileas are very easy to propagate with stem tip cuttings. Take tip cuttings in the spring. Strike 2 or 3 cuttings in a 3" pot filled with moistened potting mix. Place the pot in a propagating case or in a sealed plastic bag to hold in humidity. If kept at 75° your new plants will be rooted and growing in 3-5 weeks.

AMARYLLIS PLANT

Nothing can break the gloom, like a beautifully blooming House Plant in the dead of winter.

Of all of the flowering bulbs, the Amaryllis is the easiest to force into bloom. The large size (up to six inch blooms) of the showy flowers, as well as the ease with which they can be grown in the home make the Amaryllis extremely

popular as a gift plant during the holiday seasons

Amaryllis plants come in many beautiful colors including shades of red, white, pink, salmon and orange. There are also many bi-colored varieties, combining shades of pink or red with white.

Amaryllis is a tender bulb which originated in the Andes mountains of South America. Therefore, unfortunately, it can only be grown outdoors, year round in USDA zones 9-11. In other zones, the potted plants may be moved outdoors after the danger of frost has passed but must be brought back indoors in September.

Growing pre-planted Amaryllis

Amaryllis can be purchased as individually packaged or bulk bulbs, but more often are sold in a pre-potted condition. The pre-planted Amaryllis gift boxes normally take no more initial effort than simply watering them with room temperature water to bring them into growth.

Avoid getting water on the nose of the bulb. The plant should be kept in a brightly lighted area at about 70-75 degrees f.

After the plant begins to grow, feeding is essential. Either use a complete *slow-release* fertilizer which lasts several months, or a water soluble fertilizer twice a month. Once the plant is flowering, keep it out of direct sunlight, and slightly cooler to promote a longer flower life.

After all flowers fade, cut them off at the top of the stalk. When the stalk begins to sag, carefully cut it off just above the bulb nose. Continue to water and fertilize as normal, until the leaves begin to yellow. At this time, cut the leaves back to about 2 inches from the top of the bulb and remove it from the pot. Keep the bulb in a cool *(40-50 degree f)*, dark place for a *minimum* of 6-8 weeks.

Planting and Forcing Amaryllis Bulbs into Bloom

Any time after that, you may repot the bulb, and begin the process over again, but plant bulbs 8 weeks *before* you would like them to bloom. Bulbs may be planted at two week intervals, from September through February to ensure non-stop winter bloom, until May. Plant in a well-drained, sterile potting medium, using a standard 6-inch diameter pot.

Be sure that the pot has adequate drainage holes.

Plant the bulb, with about one-third being above the rim of the pot.

Press the soil down firmly to set the bulb securely in place.

Water thoroughly, and your plant will begin anew.

Propagating Amaryllis Plants

Bulbs that are older than two years will produce offset *bulblets*.

These bulblets may be left attached to the mother and repotted that way, creating quite a show. Preferably, the bulblets are *carefully* removed from the mother bulb *(immediately before it is replanted)*, and planted in their own individual pots. These little bulbs will take two years of growth before they will produce their first flower.

PILEAS AS HOUSE PLANTS

Pilea is a genus consisting of about 600 species of frost tender, tropical foliage plants that includes both upright bushy types and trailing varieties. Several smaller species are commonly cultivated as house plants because they are so easy to grow and care for. These inexpensive little plants are excellent for growing in dish gardens and are great starter plants for young or inexperienced growers. Their foliage varies considerably, ranging from 3" strongly textured, lance shaped leaves to tiny heart shaped, moss-like foliage. Pileas occasionally bloom, but their pink or cream colored flowers are very tiny and often go unnoticed. Growing Requirements for Pilea Plants Pileas are only hardy in USDA zones 10-12. They all have the same basic growing requirements.

Pilea plants should be grown in bright, indirect light, but never in full sun. They will *thrive* under fluorescent 'grow lights'.

Always keep your Pilea and any other house plants away from heating vents and out of drafts. Pilea plants grow best in warm daytime temperatures of 75°-80° with 10 degrees cooler at night.

Pileas need about 50% humidity in order to thrive, so they should be kept on a tray filled with moistened pebbles or misted regularly, or they can be grown in a terrarium. Keep the soil slightly moist while your Pilea is actively growing, but allow the top inch of soil to dry between waterings when growth slows in the winter. Feed monthly in the spring and summer using a half-strength solution of a liquid house plant type fertilizer with a N-P-K ratio of 3-1-2..

Pruning and Potting Pilea Plants

Pinch the tips of new growth on branching forms of Pilea to help keep the plant compact. Even with diligent pinching and regular clipping, Pileas tend to become leggy and unattractive because their lower leaves naturally drop with age. Luckily, Pilea cuttings root easily, so you can replace your old specimens with fresh new plants every year or two if you'd like. Pileas grow best in smaller, 4" pots or planters where their roots will be somewhat restricted.They should be planted in a peat moss based commercial potting mix withleaf mold and perlite added, or a mix specifically for African Violets.

Propagating Pileas with Stem Cuttings

Pileas are very easy to propagate with stem tip cuttings.

Take tip cuttings in the spring. Strike 2 or 3 cuttings in a 3" pot filled with moistened potting mix. Place the pot in a propagating case or in a sealed plastic bag to hold in humidity. If kept at 75° your new plants will be rooted and growing in 3-5 weeks.

ASPARAGUS FERNS

There are several different species and varieties of Asparagus Fern that have been grown as house plants since the late 19th century because they are are one of the fastest growing, least demanding, easiest to care for plants for growing indoors. Asparagus Ferns produce tiny white flowers in late summer to fall, and will occasionally form red or black berries. They can be trained to cascade down from a hanging planter, or to grow up a totem as a vine.

These are not true ferns but are in fact members of the Lily family.

The needle-like 'leaves' are actually called cladodes which perform the process of photosynthesis. The actual leaves are tiny and scale-like, appearing where the cladode meets the stem.

Growing Requirements of Asparagus Ferns

Asparagus Ferns are only hardy in USDA zone 9 and warmer. They all require the same *basic* care, whether they are grown in the garden or in containers, as house plants.

They grow best in bright, indirect light without any full sun. Direct, hot sunlight can scorch the 'needles', while insufficient light will often cause them to yellow or drop. They thrive in daytime temperatures of 70°-75° with it 10° cooler at night. Asparagus Ferns should be planted in rich, light, slightly acidic, well-drained soil.

The addition of peat moss to a good, commercial potting soil will take care of this requirement. While Asparagus Ferns will *tolerate* short periods of drought, they grow best when they are given plenty of water, especially when they are actively growing. In low light conditions, your Asparagus Fern will need less water, especially during the winter months. They appreciate higher humidity and an occasional misting, but this is a case where they '*appreciate*' it, more that they NEED it.

As soon as new growth begins in early spring, begin feeding your Asparagus every 3-4 weeks, using a soluble *(liquid)* all-purpose fertilizer. Continue with this feeding schedule until growth slows in the fall.

Asparagus Ferns develop large tuberous roots and can become potbound in a relatively short period of time. They will often put out more growth and flowers after they become rootbound but will dry out far quicker and will need more frequent watering or thorough drenchings. It is a good idea to re-pot established plants in the spring before new growth begins, every 3 to 4 years.

Under proper conditions, Asparagus Ferns will produce lots of vigorous, evergreen growth at all times of the year. If pruning becomes necessary, it is best to cut entire stems back to the base of the plant rather than just shortening them.

Pests and Problems that Affect Asparagus Ferns

Asparagus Ferns enjoy spending the summer outdoors in the shade of a tree, but the transition from outside to an inside environment can be quite traumatic for many plants, depending on the degree of change in light and temperature. Acclimatize them gradually from one environment to the next over a 2-3 week span.

Be sure to inspect them very carefully BEFORE bringing them back indoors, and regularly thereafter. Asparagus Ferns are very prone tospider mites, scale insects and mealybugs. If an infestation has gotten out of control, you can cut off all of the stems back to the soil line and discard the infested foliage. New stems will then grow back from the bulbs.

Asparagus ferns can develop yellow needles for many different reasons, including a change in light, rapid temperature change, over watering, under watering, and spider mites.

New growth will usually quickly resume as soon as the cause of distress is resolved.

Propagating Asparagus Ferns and Growing them from Seed

The easiest and fastest way to propagate new Asparagus Ferns is by division. The root ball can be sliced into half or quarter sections with a sharp knife or pulling the tubers apart by hand and repotting into smaller planters. Asparagus Sprengeri produces small bulblets which can be planted individually as well.

Asparagus Ferns are dioecious plants, meaning that both male and female plants must be present to produce. If your plant has berries on it, you can harvest a single seed from each one. Remove the pulp from the seed and soak it room temperature water overnight before sowing them indoors ½" deep in a tray or in pots. Maintain a temperature in the planting medium at 70°-80° until germination, which takes about 4 weeks.

BANANA PLANTS INDOORS

Planting and Caring for Banana Plants

It sounds like you are doing fine with your plant. Give it a lot of root space, and the plant will grow proportionately. Banana plants need a rich, humus soil, and heavy feeding when they are actively growing. I would use soluable 20-20-20 fertilizer monthly. Banana plants like it hot and humid so it is helpful to mist the leaves often, and sponge them off when they get dusty. Since you are growing the plant indoors, protect it from the very hottest sun in the afternoon, otherwise it needs full sun. Some varieties of banana plants produce edible fruit, others are inedible, and others even make little tiny miniature bananas. Which ever you have, it's cool!(as long as you have enough space!)

REX BEGONIAS

The Rex or Painted-Leaf Begonias are rhizomatous, clump forming, tender perennials that are only hardy in zones 10-12, so they are normally grown as house plants. Rex Begonia leaves vary considerably in texture, size, shape, patterns and colors. The small, pink or white flowers that may appear at any time of the year are insignificant compared to the foliage, but they are charming nonetheless.

Growing Requirements for Rex Begonia Plants

Rex Begonias grow best in bright, indirect sunlight coming from an east or west window. Fluorescent *grow lights* are also an excellent alternative to natural light for these plants, and in many cases will intensify the foliage colors. When potting, set the plant with the crown slightly above the soil line.

They should be planted in a porous, slightly acidic planting mix that contains some peat moss and leaf mold. African Violet potting mixes work well for Rex Begonias.

Allow the top 1" of soil to dry before watering, then water thoroughly. Overwatering can lead to root or crown rot! Wet foliage can cause problems with mildew. Always use room temperature water when watering house plants. House plants should never be allowed to stand in water for more than a few hours.

Painted-Leaf Begonia plants thrive with daytime temperatures of 70°-75° and 60°-65° at night. NEVER lower than 60°! They require a humidity level of at least 50% and appreciate periodic, light mistings.

Provide additional humidity by setting the plants on shallow trays filled with moistened pebbles, or a humidifier, especially during the winter months. Feed *actively growing plants* in early spring and again in early summer with a slow-release plant food, or every 2 weeks using a ¼ strength dilution of a soluble, complete fertilizer (23-19-14).

Decrease feeding in early fall and stop altogether in the winter. Do not over-fertilize! Your Begonia Rex may go dormant in the winter and possibly drop its leaves. If this happens, continue to water sparingly until spring when new leaves will re-appear. If your Rex is growing where it only gets light from one side *(i.e. a window sill)*, turn the pot ¼ each week to maintain even growth. Foliage color may fade under low light conditions. Consider using a grow light.

Remove any dead leaves and other debris around plant regularly. Rex Begonias enjoy spending the summer in a shaded part of the garden.

Propagating Rex Begonias and Growing them from Seed

The easiest way of propagating new Rex Begonias is with whole leaf cuttings struck in a mixture of moist peat moss and perlite. Two other ways of propagating multiple new Begonia Rex plants from a single leaf are by methods called *wedge cuttings and split vein propagation*. These methods are a bit more complicated, but they allow you to create as many as a dozen new plants from a single leaf. Maintain a temperature of 70°-75° within the growing medium until germination, which takes 15-20 days.

PARADISE PLANT

Strelitzia are only hardy in USDA zones 9-12

The Bird of Paradise *(Strelitzia reginae)* grows best when it receives a minimum of four hours of direct sunlight a day, but they should have bright light all day. Daytime temperatures of 68°-72° and night temperatures of 50° to 55° are optimal. Strelitzia flower more readily when they are somewhat rootbound, which is why they are often grown in large pots or tubs.

Plant your Bird of Paradise in a mixture of sterile potting soil, peat moss, and sand with a liberal amount of bone meal added. Good drainage is essential. Keep the soil evenly moist except from November-February when your plant should be kept on the dry side. Feed every two weeks from March through September with a balanced (10-10-10) fertilizer.

Dividing Bird of Paradise Plants

Dividing your Bird of Paradise will be easy once you have removed it from the pot. The roots are very thick and fleshy, so be careful, but don't be nervous, these plants are pretty tough. Try to separate *(unravel?)* the roots coming from each division, then carefully cut between the two plant sections with a *CLEAN, sharp knife*.

It is a good idea to dust the fresh cuts with a rooting hormone, such as 'Roottone'. Cut off any of the roots that may have been badly damaged during the operation. Repot each section in a sterile pot, using a good commercial potting soil. Keep in mind that Bird of Paradise plants bloom best when they are rootbound, so keep the new pot size as small as is feasible.

Do not water the plants for 2-3 days to give the cuts a chance to 'callous' over before resuming moderate wateringGive your Bird of Paradise a shot of all purpose fertilizer next spring, and watch them grow! Your Strelitzias should be blooming again in the following year.

Growing Bird of Paradise Plants from Seed

Bird of Paradise seeds must be sown *while they are fresh*, so plant them as soon as possible after harvestingSoak your seeds in room temperature water for 3-4 days prior to planting *(Be sure to change the water daily!)*

Plant the seeds about an inch deep and keep the potting mix moist and very warm (about 85° F.)

It doesn't matter which side of the seed is "up".

Germination can take anywhere from one month to a year, depending on the soil temperature and freshness of the seeds, *so be patient*. Bird of Paradise plants grown from seed can take seven years before producing their first bloom!

BLACK-EYED SUSAN VINES

The Black-Eyed Susan Vine is a tender, evergreen, twining vine that is most often grown as a long blooming annual. If grown as an annual, they will quickly scramble up to a height of six feet. In frost free climates they can reach 20 ft. as long they have a support to grow on.

Black-Eyed Susan Vines have dark green, arrowhead-shaped, 3" leaves. They begin blooming in the early summer and continue to produce a profusion of brightly colored, 1½" bell-shaped flowers with maroon or deep purple centers, well into the fall.

Black-Eyed Susan Vines make awesome, blooming ground covers and will quickly cover anything they can climb on, with flowers.

They grow well in hanging baskets and planters that can be brought indoors for the winter where they will continue to bloom as long as the temperature remains above 60° and they get several hours of full sun each day.

Growing Requirements for Black-Eyed Susan Vines

Black-Eyed Susan Vines are only hardy in USDA zones 9-12.

They should be grown in a spot with full sun or very light shade, except in hot summer regions where they will require shade during the afternoon. In the garden, Thunbergias should be planted in humus rich, evenly moist, well-drained soil that has been supplemented with a generous amount of compost at planting time, or in a good quality potting soil if they are grown in planters.

If you want your Black-Eyed Susan to climb, set up a trellis or other support *before* you plant!

Water regularly and thoroughly during the growing season, especially during the first year. Container grown plants must never be allowed to dry out completely. These vining plants do not tolerate drought well. A layer of mulch will help to retain soil moisture. Feed every 2-3 weeks while in bloom, using a ½ strength solution of a soluble, bloom type fertilizer *(3:1:5)*.

Any heavy pruning should be done early in the spring, before new growth begins. Black-Eyed Susan Vines can be trimmed *lightly* at any time during the growing season. Watch for scale, spider mites and whiteflies on indoor grown plants.

Propagating Black-Eyed Susans and Growing them from Seed

Black-Eyed Susan Vines can propagated by softwood cuttings taken in late spring or early summer.

New plants can also be produced by simple layering.

Pin a low growing stem to the ground, using a piece of stiff wire bent into a U, leaving the last 6 to 12 inches of the stem exposed.

Cover the pinned area with soil, then bend the tip sharply into a vertical position and nick the bark on the underside of the bend.

Provide a stake to hold it in place. Once the layer is well rooted, it can be severed from the parent and moved to another part of the garden.

Thunbergia seeds can be sown directly in the garden after all danger of frost has passed in the spring.

Start seeds indoors 6-8 weeks before the last frost date.

Maintain a temperature in the growing medium of 70°-75° until germination, which takes 10-15 days.

BOSTON FERNS

Caring for Your Boston Fern

Nephrolepis exaltata ferns like it cool, and they would be very happy to have a 60 degree room, with bright light. Since it doubtful you have a place *that cool*, I recommend that you keep your fern as far away from heat sources as possible, even at the sacrifice of some light.

Ferns thrive on humidity, so they appreciate a daily misting. If there are dead or dying fronds, cut them off, so that your plants energy can go to replacing them rather that supporting them. As far as watering, I've always had the best luck, when I would allow the pot to soak up water in a sink for a few hours, and then not water it again until the soil is dry again.

Feed your fern once a month with a half strength solution of a balanced (10-10-10) liquid houseplant fertilizer. From the sound of it you found a real treasure. I am sure that the plant is in need of some fresh soil, and to accomplish this you will either need to repot it in a larger pot, or divide it. Either of these options will create a certain amount of stress for the plant, but in the long run it will be beneficial.

Repotting Boston Ferns

If you decide to repot, remove the plant from its current pot and carefully loosen up the root mass and remove as much of the old soil as possible. The roots will have become ingrown over time, and you will want them to be able to reach out into the new soil so you will want to take a very sharp knife and make 3-4, top to bottom vertical slices, about ½" deep, into the sides of the tangled root ball. This will allow a sufficient number of roots to begin growing outward again to support the needs of your fern while it recovers. You will undoubtedly do some damage to the roots, but don't worry about it, just do your best.

You will want to use a 15 inch pot which will also be a bit deeper. Put enough good commercial potting soil into the bottom of the pot so that when you put the plant in, the original soil level will be about an inch below the rim of the pot. Clip off all of the dead fronds from the plant and put it into the new pot, then add more potting soil to the sides, watering liberally to move the soil down into the root structure.

Keep adding soil until the pot is filled to within an inch of the pot rim. The important things to remember are that you want to use good, sterile soil, and to eliminate all of the air pockets around the roots. You will probably lose some more fronds before you are done, because of shock and root damage, but just trim them off, and the plant will recover on its own.

Dividing Boston Ferns

If you would prefer to divide the plant, you basically follow the same procedure.

You can either cut the root ball vertically with a sharp butcher knife in half or quarters, or you can use your hands and seperate the roots. Cutting with the knife is easier and will do considerably less root damage.

How Grow Lights Help Plants Grow

Grow lights are special bulbs which are designed to imitate natural sun light.

Plants need both red and blue light to thrive, which they get from natural sun light but it is not available in normal light bulbs. There are several companies which produce these grow lights in florescent tubes and standard bulbs. Using artificial light will enable you to regulate the amount of time that your plants receive the light, which should be between 12 and 16 hours a day.

In some instances it is very convenient if the plant you are growing relies on seasonal changes to induce blooming, such as with Christmas or Easter Cactus. If you are growing your fern in a room with a lot of indirect, bright light you don't necessarily need the grow lights, but they sure can't hurt since the plant obviously did well under them.

BOUGAINVILLEAS

Bougainvilleas are fast growing, tropical, semi-evergreen vines from South America that produce cascading clusters of long lasting, brightly colored, petal-like bracts surrounding small tubular flowers from early spring through late summer.

In zones 10-12 Bougainvillea plants can be grown in the garden where they can be trained to climb 10-20 feet to cover porches, archways or walls or they can be grown as a colorful, spreading ground cover. Bougainvilleas are also well adapted to growing in containers and can be grown as house plants. Bougainvilleas have strong, woody thorns. The older the plant gets, the longer and bigger the thorns will be.

Growing Bougainvillea Plants in the Garden

Bougainvillea plants do not transplant well, so select a sheltered planting site in full sun where they can grow undisturbed. They aren't overly fussy about the soil as long as it is well drained. However, the addition of compost, processed manure or peat moss to the planting hole will get your Bougainvillea off to a good start.

Bougainvilleas are very drought tolerant once established and should only be watered when the soil is dry an inch below the surface, except when growing in containers which should be kept slightly moist. Feed your Bougainvillea monthly beginning in early spring using a low-nitrogen, bloom type fertilizer, following label directions carefully. Do not fertilize in the fall or winter.

Prune freely after blooming. Flowers appear on the new growth.

Growing Bougainvilleas as House Plants

Bougainvilleas can be grown as house plants but they will not bloom well indoors unless you can provide a minimum of five hours of bright sun each day and adequate humidity.

They should be grown in a rich, well-drained commercial planting mix that is NOT pre-fertilized. Repotting should be done in late winter or early in the spring. Keep the soil evenly moist during the growing season, but allow it to gradually dry out by winter. Feed *actively growing plants* every two weeks with a soluble house plant type fertilizer, following label directions carefully.

Bougainvilleas grow best with night temperatures of 65° and daytime temperatures of 70°-80°. Grow your Bougainvillea on a strong trellis or prune it in the spring to maintain a shrubby growth habit.

Container grown Bougainvilleas can be moved to a sunny spot in the garden for the summer, but should be allowed to gradually dry out before bringing them back indoors by mid autumn. Inspect your plants carefully for spider mites, mealybugs and scale before returning them to the house.

Propagating Bougainvillea Plants and Growing them from Seed

Bougainvilleas can be propagated with semi-hardwood cuttings taken in the spring. Bottom heat will speed the rooting process considerably. Bougainvillea seeds can be sown indoors at any time of the year. Maintain a temperature of 70°-75° within the growing medium until germination, which typically takes 30 days or longer.

Fig. Bougainvillea 'Purple Queen'

BRAZILIAN PLUME FLOWER

The Brazilian Plume Flower is a shade loving, tropical evergreen perennial shrub with large deep green leaves.

It produces a never ending show of large spikes of brightly colored, tubular flowers from spring to fall.

Plume Flowers are frost tender plants that are hardy in USDA Zones 9-11. If a heavy mulch is applied, they can occasionally survive in Zone 8, but will freeze back to the ground every winter.

Growing Requirements for Brazilian Plumes

Brazilian Plume Flowers should be grown in rich, evenly moist but well drained garden soil and planted where they will be in partial to full shade in the heat of the day.

Feed your Brazilian Plume plant in early spring, mid-summer and again in the early fall, using abalanced, all purpose fertilizer.

Brazilian Plume Flowers are well suited for growing in containers that can be brought into the houseduring the winter months in colder regions.

Growing Plume Flowers in the House

When growing as a House Plant, your Brazilian Plume Flower should be grown in bright indirect light during the growing season (Spring through Fall), with a room temperature of 60°-75°.

In the dormant winter months, move your plant to a cooler (55°) room where it will receive the full sun of a south facing window.

Keep the soil moist at all times but never soggy, and feed every two weeks with a good all purpose fertilizer from March through October.

Set the planter on a tray of moist stones to provide added humidity.

Pruning and Propagating Brazilian Plume Flowers

Since flowers form on the new growth of the Brazilian Plume,

pruning should be done early in the spring so flower display is not reduced.

Early in March prune your plant back hard, to about 8" or 12".

Hard pruning will help to produce a fuller, multi stemmed plant.

Brazilian Plume Flowers are very easy to propagate with 8"-12" softwood cuttings or 2"-3" tip cuttings.

Remove all but the top 2 leaves and dust the root end of the stems with rooting hormone powder before striking them in moist potting soil.

Plume Flowers can be started indoors from seeds.

Justicia seeds require light for germination, so do not cover them up!

Maintain a temperature of 70° within the growing medium.

Germination takes 20-25 days.

Fig. Brazilian Plume Flower Justinia carnea

CARNIVOROUS BUTTERWORT PLANT

The plants themselves are fairly plain in appearance. Their name is derived by the buttery, sticky, greasy mucilage which covers their leaves. Their extremely pretty flowers, however, usually make up for these shortcomings.

The *Pinguicula* family of plants capture their prey using a fly paper approach. Insects that are attracted to the plant become stuck in the leaf's sticky coating. Then they are digested through the action of the enzymes in the plant's slime, and become nutrients for the plant.

On certain varieties of Butterworts, the leaf margins slowly fold over and entrap the insect before the digestive process begins.

Butterwort plants are suitable for growing in unsealed terarriums. Growing Requirements of Butterwort Plants

- Butterwort plants should be grown in a mixture of sphagnum moss, peat moss and perlite (or coarse builders sand).
- Butterworts grow best when they receive several hours of sunlight each day.
- Butterwort plants will tolerate a wide range of room temperatures (60-90 degrees) as a house plant.
- Butterworts will naturally go into a dormant period *(like many of the sundews)*. When this happens, the plant will take on a new appearance, forming many new, small leaves around the head called a hibernaculum.
- Butterwort plants require a varying amount of moisture, depending on the species, so investigate the needs of your particular plant.

Pinguicula planifolia requires very wet conditions, and can even survive submerged in water for extended periods of time.

Pinguicula caerulea and *P. lutea* are most often found on *drier* ground, but should never be allowed to completely 'dry' out.

Fig. Butterwort Plants

CALADIUM PLANTS

The Caladium is a frost tender, perennial tuber which is grown for its colorful foliage rather than flowers. The leaves are usually a combination of different shades of red, pink, green, white.

Growing Requirements for Caladium Plants

Caladiums are easy to grow plants, but they tend to be fussy about their growing conditions.

Primarily, they like heat, plenty of moisture and relatively high humidity. Most varieties of Caladium prefer to grow in a partially shaded area of the garden, but will tolerate morning sun. Caladiums are hardy in USDA zones 10-11.

In warm, frost free climates, Caladiums can be planted directly in the garden where they can be left permanently. In cool regions they should be potted and started indoors. When the soil temperatures rises to 70 degrees, move the plant directly into the garden.

An alternative method is to sink the pot up to the rim in the garden for easy removal prior to the first frost. Caladiums can also be grown in containers, as house plants. In the house, keep them in a bright *(not sunny)* location, or grow them under artificial 'Grow Lights'.

How to Plant Caladium Tubers

The knobby side of Caladium tuber should be planted upward.

The leaves emerge from the eyes (knobs) and the roots grow from the top of the tuber. (Don't worry if you planted the tuber upside down, Caladiums will orient themselves and grow to the surface anyway, but they will take a little longer to show)

Plant the tuber in rich, well draining soil so that the top of it is two inches below the surface.Provide plenty of water and warmth, and the sprouts will begin to emerge in 4-6 weeks.

Feed lightly every 6 weeks when actively growing, using a liquid 6-6-6 fertilizer.In cool regions, Caladium tubers should be dug before the first frost.

When night time temperatures start to drop near 50 degrees, cut back on watering to allow the leaves to dry up and fall off naturally.If necessary, the leaves can also be removed by cutting them off 1/2 inch from the tuber.

Store the tubers in a flat of peat moss or vermiculite in a warm, well ventilated area.Check periodically for softness or rot and discard any bad tubers.

Fig. Caladium Colocasia esculenta

If the bulbs begin to shrivel, add a very small amount of water to the storage medium. Replant them again in the spring. Caladium tubers can be cut with a sterile knife to produce new plants. For best results, each cutting should contain at least two 'eyes'.

CHILEAN MITRE FLOWER

Mitraria coccinea

The Chilean Mitre Flower is a gorgeous creeping and climbing, loosely branching, evergreen sub-shrub from the cool rainforests of Chile. Mitre Flower plants have glossy, dark green, 1"-1½" toothed leaves arranged in pairs along woody stems that can easily reach 6 feet.

From late spring until fall they produce an abundance of intensely colored, scarlet-orange, tubular 1½" flowers with a soft, velvety texture. The flowers are followed by glossy red, ½" berry-like fruit.

Chilean Mitre Flowers can be allowed to creep along the ground, climb on a fence or trellis, or just have them cascade down from a hanging basket.

Growing Mitre Flowers as House Plants

Chilean Mitre Flowers make excellent house plants, provided that they receive bright, indirect light and are grown in a cool *(65°-70°)* room. They should should be planted in a quality potting mix that has a high peat moss content with perlite added for drainage.

Mitre Flowers resent transplanting, but when it becomes necessary, water the plant thoroughly before starting, and take care not to disturb the roots any more than is absolutely necessary.

Keep the soil evenly moist, but never soggy. City water is treated with chemicals for your safety, however plants don't like chlorine *or* fluoride, so it's a very good idea to allow the water to sit in an open container (un-capped milk bottle) for at least 24 hours prior to using it on your house plants.

This also allows the water to warm up to room temperature.

Feed your Mitre Flower every 6-8 weeks when it is actively growing, using a good, soluble fertilizer that has been formulated specifically for *flowering house plants*. Always follow label directions!

Remove spent flowers and dead foliage regularly. Prune as needed to keep the plant tidy.Watch out for spider mites and whiteflies.

Growing Requirements for Chilean Mitre Flowers

Chilean Mitre Flower plants are only hardy in USDA zones 9-11. They are happiest where temperatures remain below 85°, with cooler nights. Ideally, they should be grown where they will receive sun in the morning and partial shade during the heat of the afternoon. Plant them in evenly moist, slightly acidic but well-drained soil that has been enriched with a generous amount of well decayed compost at planting time.

Feed every two months when actively growing, with a Rhododendron type fertilizer. Always follow label directions!

Take precautions to protect your Mitre Flower from slugs and snails.

Watch out for spider mites, ants and whiteflies.

Propagating Chilean Mitre Flowers and Growing them from Seed

Mitre Flowers are easily propagated with 6"-8" tip cuttings taken in early summer and struck in a moist mixture of peat moss and perlite.

Another easy way of propagating new Mitre Flower plants is by simple layering. Using a piece of stiff wire bent into a U, pin a stem to the ground about 6"-12" from the tip. Cover the pinned area with moist soil. Once the layering is well rooted it can be severed from the parent and replanted elsewhere in the garden.

Mitre Flower seeds have a very low germination rate of about 25%. Seeds should be sown directly in the garden in late fall, while they are still fresh.

Fig. Chilean Mitre Flower Mitraria coccinea

CHRISTMAS CACTUS PLANTS

Schlumbergera

Schlumbergera plants are thermo-photoperiodic.

That means that their *bloom set* timing is triggered by a combination of day length and temperature.

The most important trigger is the length of the day.

Once the day length decreases to approximately twelve hours, the plant recognizes that it is time to bloom.

Night temperature of 55° to 65° F. should accompany the short day.

There are six species of Schlumbergera that are all native to southwestern Brazil. The genus was named after Frederic Schlumberger, a 19th century French cactus collector who discovered the *original Christmas Cactus*, Schlumbergera buckleyi.

Most Christmas Cactus sold today are hybrids of this red flowering species.

Schlumbergera are epiphytic plants with flattened, 2" leaf-like stem segments. In nature, they reside cordially with the Orchids and Epiphyllums in the crotches and branches of their host jungle trees.

Look at the calendar and you decide what you want to call your Holiday Cactus because Schlumbergera plants are capable of blooming at almost any time of the year, by manipulating their light and temperature. The natural changes of daylight hours will usually trigger blooming of these different Holiday Cactus to coincide with their named holiday.

Growing Requirements for Christmas Cactus

Like most epiphytic cacti, Christmas Cactus like water but don't like to be soaked. They can grow with very little light and that's why they make good house plants, however, for good flower production your Christmas Cactus will require bright light or full sun in the fall months.

Plant them in a soil mix consisting of 3 parts peat moss, 1 part loam and 1 part coarse sand or perlite.

Schlumbergeras bloom best when they are *somewhat* rootbound and crowded. Plant them in a pot that is wider than it is tall.

When repotting, only move up one size of planter.

The soil should be kept moist during the spring and summer. Fertilize every 2 weeksduring this time with a diluted, ½ strength houseplant fertilizer.

In the fall, reduce watering and keep your Christmas Cactus cooler 60°-65° at night. Protect from artificial light during night time hours.

Increase watering and resume feeding when the flower buds appear. Christmas Cactus and other Holiday Cactus do not like to be moved to a different location once they set their buds.

Environmental changes can cause the flowers to rapidly drop.

Propagating Thanksgiving and Christmas Cactus

Schlumbergera are easily propagated by cuttings taken in the spring or summer.

Cut (or twist off) single or multiple segments from the tips of the branches.

Keep track of which end is 'up'.

Set the segments in a cool, dry location for a week to let the cuts callous over.

Insert the base of the cutting ½" deep in moist sand or vermiculite mix.

When the new roots are well established, move your new Christmas Cactus to an appropriate container

Fig. Thanksgiving Cactus

Schlumbergera truncata

The Thanksgiving Cactus (front right in first photo), was originally named Zygocactus truncatus, and sometimes goes by the name of the Crab Cactus.

The most distinguishing trait of S.truncata is the claw toothed edges on each segment. Flowers may be pink, red, orange or white.

Christmas Cactus

Schlumbergera x buckleyi

Christmas Cactus *(white, top center in first photo)* have a much less pronounced 'claw' on the edges of the leaf segments.

The majority of Christmas Cactus on the market today are hybrids of the original, red flowering *Schlumbergera buckleyi*.

Flowers may be pink, red, orange or white or lavender.

CARNIVOROUS COBRA LILY PLANTS

Darlingtonia californica

The Cobra Lily, *Darlingtonia californica*, is native to southwestern Oregon and northern California.

This carnivorous plant lures it's insect prey with a sweet nectar which is inside the leaf opening under it's hood.

Once inside, the insect becomes *confused* by the many transparent areas of the upper leaf surfaces, which appear to be exits.

As the insect checks these false exits searching for an escape route, it is led down the tube structure and is unable to return to the top of the plant because of the slippery smooth surface of the inner tube and the sharp, downward pointing hairs which effectively block any chance of escape.

Eventually, the insects will fall into a pool of liquid digestive enzyme in the base of the leaf where they are absorbed as food for the plant.

Growing Requirements for Cobra Lily Plants

Cobra Lilies like bright light but not direct sunlight. *Ideally,* they should be grown in living sphagnum moss, but can be grown in a mixture of peat moss and perlite if live sphagnum is not available.

High humidity is not nearly as important for these plants as it is for certain other carnivorous plants, but a certain amount of humidity should be maintained. Although Cobra Lilies can *withstand* temperatures as low as freezing or as high as 100° F., the optimum growing temperature range is between 70° and 80° F., with night temperatures being about 20 degrees cooler.

Watering your Cobra Lily

It is very important to keep the root system cool.

Watering is by far the most important aspect of keeping your Cobra Lily healthy. It's natural habitat is in boggy areas, near mountain streams where it stays relatively cool and constantly has cool water running over the root system.

Duplicating this natural environment as closely as possible is extremely important to it's cultivation as a house plant.

This can be best accomplished by flushing fresh water (distilled or rain water) through it's pot every day. This simulates the cool water that is constantly running over their roots in their natural habitat.

A Cobra Lily may grow to four feet tall in cultivation, so eventually you will need to find a much larger container to use.

As the plant matures it will probably produce a great number of side shoots and eventually form a large clump.

It is not a good idea to separate the clumps because of the sensitivity of the root system.

Once divided, the plant may go into shock and fail to grow for an extended period of time, and will be susceptible to disease and fungal attacks.

Cobra Lilies are fairly hardy, and in nature can survive being buried in snow during the winter.

This makes them an excellent addition to an outdoor bog garden in USDA hardiness zones 7-10.

Before you leave please read How to Grow and Care for Carnivorous Insectivores.

Fig. Cobra Lilies Darlingtonia californica

COFFEE BEAN TREE!

Coffea arabica

That is how long it will take you before you can harvest your first beans from your own coffee plant, *(Coffea arabica).* In the meantime, you can enjoy the dark shiny green leaves and fragrant white flowers of this very unique, ornamental plant.

Growing your own Coffee Tree Plant

The coffee 'tree' is actually a variety of tropical evergreen shrub.

Coffee trees do not begin to produce its full yield until its sixth year and will continue prime production for about ten years, however Coffee plants may live on for 60 years. The tree, if left alone will grow to a height of between 16 and 40 feet. In most coffee plantations the trees are kept at a manageable six feet to get the best yield and to make it easier to harvest. The best growing conditions are in a temperature range of 65 degrees Fahrenheit to 75 degrees Fahrenheit. Rainfall should be plentiful and the weather should switch between heavy rainfall and sunshine to bring the berries to full maturity. The type of soil is not too important but good drainage is a must!

Coffea plants are fairly easy to grow in the home as a potted specimen, but they should be moved outdoors for the summer if possible. They grow best in filtered sunlight, with night temperatures in the lower to mid 60s and day temperatures of 70F or higher. Plant them in any good commercial, fast draining potting soil. The soil should be kept on the moist side, but never soggy.

Coffee tree plants will produce fruit without any fertilizing whatsoever, but for best results and maximum yield, they should be fed every 2 weeks from March to October, and then monthly from November through February. Use a soluble, all purpose *(10-10-10)* fertilizer.

Coffea bean plants are easily propagated from seeds. You can find seeds for coffea and other rare and unusual plants online from *The Whatcom Seed Company*

Coffee trees will grow to about 10 feet if given ample root room, but can be pruned if this is too big for the allotted space. Pruning may involve simple pinching to produce a bushier plant, or you may go as far as cutting it way back.... Right down to where only two branches are left near the bottom of the trunk.

Harvesting your Coffee Beans

The fruit are red when they ripen in the fall, with a sweet pulp surrounding the bean. Each coffee berry produces two beans. The coffee tree's fruit does not all ripen at one time. In fact it will have blossoms and berries in various stages of ripening. Only the ripe berries can be picked. The berries cannot be picked when green since they will not ripen once picked.

Roasting Coffee Beans

Once you have harvested sufficient beans to brew your first pot of 'home grown' coffee, you will have to roast them. There are many 'home type' roasters available on the market, which do an excellent job of evenly roasting your beans. Whether you are willing to go to the expense of purchasing one of these is up to you. It is possible to roast your own beans in the oven. This method will tend to smoke up the house a bit, and the smell of the burned off chaff will tend to linger in the house for quite some time.

The amount and size of beans, as well as your altitude will make a difference in the roasting process, so this is a 'live and learn process. The beans should be placed in some form of perforated container, such as a steel strainer or vegetable steamer. Place them in a pre-heated, 250 degree F. oven for about seven minutes. Then increase the oven temperature to 450 degrees. In about ten minutes the beans should begin to crackle. (This timing will depend on the temperature, heat capacity of the oven, and the beans; so keep an eye on them.) When the beans start crackling, mix them up, to promote an even roast. Check their color every two minutes until they have achieved a color slightly lighter than the end color which you desire. *(As the beans cool, they will continue roasting).*

When they have completely cooled, store them in an air tight container in the refrigerator.

Fig. Coffee Bean Tree Coffea arabica

CROTON PLANTS

Codiaeum variegatum

Crotons are tender perennial plants that will grow to three feet tall or more.

They are most often grown as house plants or in a greenhouse environment.

Their thick, glossy foliage is brightly colored in combinations and shades of red, yellow, pink, orange, burgundy, bronze or green.

The leaves may be wide and smooth, long and narrow, or very irregularly shaped.

Growing Requirements for Croton Plants

Unfortunately, Crotons tend to be pretty fussy about their growing environment.

Low humidity, drafts, drought, temperatures below 50°, or rapid changes in temperature can all cause the leaves to drop.

In USDA zones 11 and 12, they can be grown in the garden or in containers in a space where the will get full sun.

Crotons can be grown in partial shade, but will lose much of their bright coloring.

Larger growing cultivars make excellent hedge plantings.

Narrow leaved varieties of Croton are considerably easier to grow and care for.

Growing Crotons as House Plants

When growing your Croton as a house plant, it should be planted in a rich, commercial potting mix. Grow them in a draft free area where they receive full sun or very bright light for at least 4 hours a day.

Night time temperatures should be about 65° and a daytime temperature range of 70°-80°, with 80° being ideal.

Crotons require high humidity (near 70%). Unfortunately, when temperatures in the home rise above 67°, the humidity drops drastically, so it may be necessary to sacrifice a few degrees of warmth in lieu of an increase in the humidity.

Additional humidity can be created by setting the planter on a tray filled with moist pebbles, creating an indoor greenhouse or placing an aquarium nearby. Keep the soil evenly moist except in December and January when you should allow the soil to dry slightly to allow for dormancy.

Feed Crotons monthly when they are actively growing with a fertilizer specifically formulated for house plants.

Propagating Croton Plants and Growing Them from Seed

Crotons are easily propagated with tip cuttings taken from new growth in the spring or summer and rooted at 70°-80°.

Start Croton seeds indoors at any time of the year.

Maintain a temperature of 75°-80° within the growing medium until germination, which takes about 30 days.

Fig. Croton Codiaeum variegatum

CROWN OF THORNS PLANT

Euphorbia milii

The Crown of Thorns is a woody, multi-stemmed, succulent shrub that originated in Madagascar. Dark green, tear shaped leaves appear randomly on each thorn covered branch.

When a Crown of Thorns plant becomes stressed due to over watering or under watering, humidity or temperature changes, it may quickly drop all of its leaves. Normally, once the cause of stress is resolved, fresh foliage will quickly return to your plant.

Crown of Thorns will produce flowers nearly all year, but especially during the winter months. The tiny flowers are insignificant but appear just above brightly colored red or yellow bracts.

A Crown of Thorns Plant in Distress

"My crown of thorns has developed some problems over the past several months.

As background, I received the plant from my mom about 7 years ago (she's had it for about 20 years).

I think it has bloomed, maybe, 3 or 4 times since I've had it.

It was in fairly good shape (despite the lack of blooming) until I moved to NY 6 months ago. Before I moved I think I repotted it once in 6 1/2 years, it got plenty of sun and I generally ignored it, only watering it when the leaves would look dry and overly curly.

Now the plant is not very full, the leaves get brown and yellow spots on them, proceed to dry up, curl over, turn yellow and fall off with alarming regularity and speed. Several leaves fall off every day. The plant isn't nearly as full as it once was and the spines on the stem are brittle and break when you touch them instead of poking you.

Growing Requirements for Crown of Thorns Plants

Since the downhill slide of your Crown of Thorns began about the time of your move, my first guess would be that the heating is different in your new home resulting in drier air.

Because of their leaves, most people don't consider a Crown of Thorns to be a succulent plant, but in reality it is. They have limited water storing abilities, however, and often have to rely on moisture it takes in through its leaves so I would try misting your plant daily. You might also want to increase the amount of water you give it each time you water it.

Crown of Thorns plants should be watered once a week when it is in a growth cycle, but only once a month while dormant. Whether dormant or actively growing, *never water before the soil has completely dried.*

Once you have resolved the leaf drop issue, you should start feeding your plant monthly with 'half-strength' liquid house plant BLOOM fertilizer to get your plant blooming again. Crown of Thorns plants should have some direct sun each day, even if it just an hour or two.

Propagating a Crown of Thorns Plant

CAUTION! The white milky sap is VERY poisonous, use gloves and wash your hands thoroughly afterwards! First of all, you need to cut any of the rotted branches back to a point where you find clean healthy growth.

If the rot started in the roots you may only be able to salvage a lot of cuttings, however if the rot is only at the branch tip, you can cut it back to healthy wood, and your plant will come back from the rootstock. It is possible to take and root cuttings from a Crown of Thorns plant but they take a little more effort and time than most cuttings.

Using a very sharp, clean knife *(I like single edge razor blades)* cut a branch off at the point where it meets the main trunk. Set the cuttings aside in a cool dry place for two or three days to allow the cuts to callous over. Once the callous has formed, dip 1-2 inches of the cut end into a rooting hormone, such as Roottone®. Insert each cutting into a clean pot of sterile, sandy potting soil. The pot should then put in a warm place with bright light. Do not water at all for the first 2 weeks, and then you can begin watering the cuttings *very sparingly* until you see signs of new growth *(usually in about a month)* Rooting will be faster if the soil is kept at a constant 75° F. I suggest that you take several cuttings to be sure that you succeed with at least one of them.

CYCLAMEN PLANT

The blossoms of the Cyclamen twist and turn like tiny butterfly wings.

The silvery, marbled coloring of the foliage accents the beautiful flower colors of lavender, rose, maroon, red, or white.

Growing Cyclamen as House Plants

The most important criteria for success with growing a Cyclamen are cool temperatures, fresh air, and ample moisture. Cyclamen will rarely succeed if the temperature goes above sixty-five degrees during the day and fifty degrees at night. Cyclamen said should be watered daily to keep the soil moist. Because they prefer a humid atmosphere, misting is beneficial. Ideally, they should be kept on top of a pebble fill tray of water to provide added humidity. Ideally, they should be watered with room temperature, but you should *never* pour the water directly over the crown of the plant, as this can cause the corm to rot.

Dormant Period for Cyclamen

Your Cyclamen will continue to bloom from December up until April if it is kept cool and happy. As the flower stalks and leaves die off they should be removed by giving them a sharp tug. New foliage and flowers will soon replace the old. The corm of the Cyclamen plant will naturally go into dormancy sometime after April. At that time, gradually reduce the water and keep the plant in a cool, darker place.

Allow the corm to slowly dry out. Replant it in a shaded area of the garden in the spring. In the garden, Cyclamen should be planted with half of the corm just below the soil surface. In late July or August, when the tiny new leaves begin to develop, transplant it into a pot using a rich sandy potting soil but leave it outdoors.

Keep the bed on which the pot stands moist. In early fall the potted corm can be again moved into the house, where it should be kept in a window where it will receive a few hours of *morning* sun each day.

Propagating Cyclamen Plants

Cyclamen can be grown from seed, but it takes up to eighteen months before you will see your first flowers. You may be better off to purchase your plant from a nursery if you tend to be an impatient gardener.

Growing Hardy Cyclamens in the Garden

Hardy Cyclamen should be planted in a semi shady area of the garden in August or September. The corms should be planted 2 in. deep and 6 in. apart in a slightly alkaline soil, with a generous amount of leaf mold added.

Different varieties of hardy Cyclamen may root from either the bottom or top of the corm. Be certain to plant the appropriate side up!

Fig. Cyclamens in Bloom *Cyclamen persicum*

DIEFFENBACHIA PLANTS

The Dumb Cane Plant

"I received this plant approximately 4 months ago at my grandfather's funeral.

Since that time, it has not seemed to flourish. I have tried to water it frequently, keep in sunlight when possible and also kept the temperature inside at approx. 65 - 75 degrees. It has begun to thin out, droop, leaves turn brown.

I do not know what to do for this plant. Am I doing all the wrong things?

I am not a plant person, but I would really like to save this one since it was given to me in my grandfather's memory.

I have not changed the pot it is in. Do you think it needs to be put in a bigger pot?

Thank you for your help."

Warning!!! The sap of the Dieffenbachia plant is extremely poisonous!

Even a small amount in the mouth can cause the tongue to swell to the point that it will close your throat and cause suffocation. Contact with the skin can cause severe irritation or rashes. Wash your hands thoroughly after you have any contact with the sap!

Growing Requirements for Dieffenbachia Plants

Dieffenbachias grow best in the bright *indirect* light of north or east window. The ideal growing temperatures are between 70° to 80° F. during the day and 60° to 65° at night.

Dieffenbachia plants will perish if kept below 50°!

The soil should be allowed to become moderately dry between waterings, and then the plant should receive a *thorough* drenching.

Established plants should be fed with every two or three weeks during the spring and summer with any all purpose house-plant fertilizer diluted to one half strength.

Dieffenbachia Growing Tips

Root bound Dieffenbachias can be repotted at any time of the year, using any good commercial potting soil.

The leaves of your plant should be cleaned occasionally with a damp sponge to prevent the pores from being clogged with dust.

As your Dieffenbachia gets larger, the lower leaves naturally wither and die, creating the distinguishing 'cane'.

This natural function allows the plant's energy to be diverted to the production of new top leaves. If the plant becomes to tall and leggy, it can be cut back to any height and the cut portions used for propagation.

The uncut portion of the cane will then begin to branch from the old leaf nodes.

Fig. A Dumb Cane Plant Dieffenbachia picta

MADAGASCAR DRAGON TREE

Dracaena marginata

Dracaena marginata is a very durable, easy to care for plant that will slowly grow to a multi stemmed, 10 feet tall shrub.

It has a flexible, snake-like trunk that will eventually branch with age, or can be forced to branch by cutting the stem tip.

The Madagascar Dragon Tree has narrow, dark green, arching leaves that are edged in deep red.

Many hybrids have been introduced that have other colors of foliage, as well.

In most areas the Dragon Tree is grown as a house plant, but for those who are lucky enough to grow their gardens in frost free areas of USDA zones 10-12,

Dragon Trees make an excellent, drought resistant, accent plant for growing in partial to light shade.

Growing Dracaena marginata as House Plants

Dragon Trees will adapt to a wide variety of light conditions, but will have the best foliage color when they are grown in bright, indirect light.

Green varieties can survive with considerably less light.

They should be potted in a well draining house plant soil.

Dracaena marginata plants are very drought resistant, but they should be watered on a regular basis, nonetheless. Wait until the soil surface is dry to the touch, then water them thoroughly, using water that has been allowed to set for 24 hours.

Never allow the soil to remain constantly wet!

Dragon Trees appreciate an occasional misting of their foliage.

Fertilize them monthly with a balanced house plant fertilizer.

During the winter months, fertilizing should be withheld, watering reduced and your Dracaena should be moved to a cooler room.

Watch out for mealybugs, scale insects and mites, as they are the most common insect pests that attack Dracaena plants.

Propagating Dragon Tree Plants

Dracaena marginata are easy to propagate by air layering, or with tip or stem cuttings taken in late spring or summer.

Fig. Dragon Tree Dracaena marginata 'tricolor'

EASTER CACTUS PLANTS

Rhipsalidopsis gaertneri and Rhipsalidopsis rosea

Like the Christmas and Thanksgiving Cactus, Easter *or* Spring Cactus, as they are also known, are epiphytic plants that are native to the tropical jungles of Brazil.

The Dwarf Easter Cactus, *Rhipsalidopsis rosea* produces soft pink, 1"-2" flowers.

Rhipsalidopsis gaertneri produces bright red flowers that can reach 3".

This species of Easter Cactus has been vigorously hybridized with *R. rosea* and various *Epiphyllums* to create many new varieties in colors ranging from pure white to bright purple.

Both species of Easter Cactus bloom early in the spring, and will occasionally flower again in the fall.

Growing Requirements for Spring and Dwarf Easter Cactus

Spring and Dwarf Easter Cactus are frost tender plants that are only hardy in USDA zones 9-12, but they are reasonably easy to grow and maintain as house plants.

As with growing most other house plants, Easter Cactus grow best in an environment that is close to their native habitat... Fortunately, however, these plants are quite willing to adapt.

In the garden, Easter Cactus should be grown in planters or hanging baskets in dappled to full shade.

They will tolerate some morning or late afternoon sun in coastal or temperate regions.

Indoors, these plants should be grown in bright, indirect light, but never in full sun, in an area with good air circulation.

Easter Cactus can *survive* in temperatures from 35° to 100°, but to *thrive*, they need daytime temperatures of 70°-80° with 10° cooler at night.

Planting and Repotting Easter Cactus Plants

Excellent drainage is essential for both the planter and the planting mixfor your Easter Cactus to survive.

Easter Cactus should be grown in a light, fast draining, peat moss based planting mix with coarse builders sand or perlite added.

Water actively growing Easter Cactus thoroughly when the top inch of soil becomes dry to the touch.

Never allow the pot stand in water for more than an hour. Waterlogged soil can quickly lead to root rot and disease.

Established, healthy Easter Cactus plants should be fed monthly beginning in late February until September, using a soluble fertilizer with an NPK ratio of 10-10-5.

Rhipsalidopsis plants will become semi-dormant in the fall and winter months, while they are setting new buds.

During this time, they should be kept in a cooler spot (50°-60°) and watering should be cut back to once or twice a month. Easter Cactus bloom best when they are *somewhat* rootbound and crowded, but they should be repotted in fresh potting mix every other year to a pot that is one size larger. Repotting should be done in mid-summer after blooming or in the fall.

Easter Cactus are prone to attacks from slugs and snails, spider mites and mealy bugs.

Propagating Easter Cactus Plants

Easter Cactus can be easily propagated by cuttings of segments taken during the summer after your plant has finished blooming.

Cut *or gently twist off* single or multiple segments from the tips of the healthy branches. *Be sure to keep track of which end of the cutting is 'up'.*

Set the segments in a cool, dry location for several days, to allow the cut ends to form a callous. Insert the base of the cutting ½" deep in moist sand, perlite or vermiculite mix.

When the new roots are well established, move your new Easter Cactus to an appropriate container in a rich, fast draining potting mix.

Fig. Spring Cactus Rhipsalidopsis gaertneri

Fig. Dwarf Easter Cactus Rhipsalidopsis rosea

ELEPHANT EAR PHILODENDRON

Philodendron bipinnatifidum

April 3,1998 Mandy McDevitt wrote: Hi. My name is Mandy and I have a question regarding a 15 year old Philodendron ("elephant ear") that I've potted and repotted five times over that stretch. I'm at the point now where I can't separate the babes from the original, so I have about 9 good sized plants in one 2' x 3' pot. It goes out on the deck from spring to fall and stays in during the winter.

It's so healthy! But I'd like to separate it... if just to give the parts of it a rest! But also to separate out some of the babies and put them in other places around the house. Of course we don't have any room for this but we make do! (smile) How do I separate these fused babies; I mean the whole thing comes out of the pot in a clump - hardly any divisions can be seen. Five or six years ago I split it with a hatchet (yes, and I'm ashamed) but managed only to cut a wedge out of the bottom. I put the clump back in a bigger pot, covered it over and it's thrived (and given birth) ever since. Should I take a saw to it now? I apologize if this is making you shudder. Just throwing out thoughts. Good luck with this. I'll be very interested in what you might suggest. and thanks... Mandy

Well, Mandy sometimes you just have to do what ya gotta do! The Elephant Ear Philodendron or Selloum Philodendron, *(Philodendron selloum)* is very tough, so it can take a lot of abuse, and end up looking better in the end. I only hope that when you took the hatchet to it, you didn't have some sort of evil, crazed smile on your face. If I were taking on your project, (and I have done it *many* times). I'd go about it this way.....Try to think of yourself as a surgeon, use *clean,* sharp tools First I would slide the whole thing out of the pot, and lay it on its side on a bench, or the kitchen table. Usually, with a little patience you can find one or two of the pups which you can remove from the parent plant. If you do, just try to follow the roots into the ball, untangling as you go. If you happen to break a root or two, just clip them to a nice clean cut. The soil in the root mass is more than likely old and tired, so clean as much of it away as possible.

Then I would use a sharp serrated knife, or hand saw to cut through the root ball, and then separating the plant portion as much as possible. I wouldn't cut the root ball into any more than quarters, myself....Make sure that all of the cut roots have been cleanly cut, and then dust the cuts with a rooting hormone such as*Roottone.* Allow the roots to remain exposed for a couple hours to allow the cuts to callous over before repotting with a good commercial potting soil. Don't water at all the first day, but then water very well, using the force of the water to wash down additional dirt down into the root zone. You will have to add more potting soil to replace what you have removed. Your new plants will go through quite a bit of shock in this procedure, so it'll take a while before they're looking great again. Hold off feeding until the plant has started to show signs of new growth.

FALSE ARALIA PLANTS

Schefflera elegantissima fka: Dizygotheca elegantissima

False Aralias are small, tropical evergreen trees that are native to the islands of the South Pacific.

In their native environment, *Schefflera elegantissima* can grow up to 26 feet tall, but when grown in pots or planters they seldom reach more than 4-6 feet tall. They are single stemmed plants withpalmate leaves that are divided into 4-9 dark, greenish-brown leaflets with deeply serrated edges.

Schefflera veitchii is a similar species, but has coppery-green leaves with light red veins.

False Aralia plants seldom bloom in cultivation.

They are fairly easy to grow as house plants in areas where they are not hardy.

Small plants are an excellent choice for growing in dish gardens and in terrariums.

Growing Requirements for False Aralia Plants

False Aralias are only hardy in USDA zones 11-12. In other areas, they are grown as house plants.

In the garden, *Schefflera elegantissima* should be grown in partial to full shade where they will be protected from the wind.They should be planted in slightly acidic, well drained but moisture retentive soil with a pH of 5.5 to 6.5.

Water regularly to keep the soil evenly moist, but never soggy.

Feed False Aralia plants monthly, using a balanced, all purpose fertilizer.

Growing False Aralias as House Plants

Indoors, False Aralia plants should be kept in an area with bright filtered light, but never in full sun.

They should be grown in pots or planters with sufficient drainage holes, and be planted in a peat moss based, commercial potting mix.

Water regularly while actively growing, keeping the soil slightly moist.

Never allow the soil to remain soggy nor let it dry out completely. Never allow a False Aralia to sit in a saucer of water.

False Aralias prefer a warm, fairly humid atmosphere, with daytime temperatures of 75°-85° and 70° nights.

Temperatures below 60° will quickly cause the lower leaves to drop.

Feed pot grown False Aralia plants every other month during the spring and summer, using a soluble fertilizer that has been formulated for house plants.

Your plant will enjoy spending the summer months in a shaded area of your garden, but be sure to check for pests and problems before bringing them back indoors.

During the dry winter months, you should mist the foliage daily or place the planter on a tray filled with moistened pebbles to increase humidity.

Pests and Problems of False Aralia Plants

The most common pests to infest False Aralias are Spider Mites, Aphids, Scale and Mealy Bugs.

False Aralia plants will shed their lower leaves for a variety of reasons.

Watering is critical. Too much or too little water will cause the leaves to shed.

The leaves will also fall off if the temperature drops below 60° or changes rapidly.

Lack of humidity in the winter is another cause of leaf drop.

Propagating Schefflera elegantissima Plants

False Aralia plants can be propagated with softwood stem cuttings taken during the summer. Older plants are a little harder to start and should be propagated by air layering in the spring.

Schefflera seeds can be started indoors at any time, but for the best results sow *fresh* seeds in late summer.

These seeds germinated faster with light, so do not cover them!

Maintain a temperature in the growing medium of 70° until germination, which takes 20-30 days.

Fig. False Aralia Schefflera elegantissima

FIRECRACKER FLOWER PLANTS

Crossandra infundibuliformis

Firecracker Flower plants are very decorative, compact flowering shrubs that grow 18"-24" tall. They have glossy, dark green leaves and spikes of brightly colored, funnel shaped flowers that continue to bloom for months on end.

Firecracker Flowers are easy plants to grow and care for, as long as they have sufficient water and humidity. They are excellent plants for growing in planters which can be brought indoors in the winter months and grown as a house plant.

Firecracker Plants will not tolerate temperatures below 55°

Growing Requirements for Firecracker Flower Plants

Unfortunately, Firecracker Flowers are only hardy in USDA zones 10-12.

When grown outdoors, Firecracker Plants should be grown in partial shade to full sun, and planted in well-drained soil that has been amended by adding an equal part of peat moss.

Indoors, keep your Firecracker Flower in an uncrowded, draft free area where it will receive bright, indirect light, with a little morning sun if possible. They grow extremely well under fluorescent grow lights. Firecracker Flowers thrive in warm, daytime temperatures of 75° with cooler, 60°-65° nights. They require relatively high humidity, so it is a good idea to set the planter on top of a tray filled with moistened stones. Misting the foliage is helpful, but do not mist the flowers!

During the active growing season from March to October, the soil should be kept evenly moist but never soggy. Feed every 2 weeks with a soluble, house plant fertilizer that has been diluted by 50%.

Reduce watering and stop feeding during the dormant winter months.

Repotting should be done in February, using a rich, fast draining, peat moss based potting mix.

Cut back stems that have flowered by one half in late winter to promote branching. Keep a vigilant watch out for for spider mites and white flies.

Propagating Firecracker Plants and Growing them from Seed

Firecracker Flower plants are easily propagated with 3"-4" stem cuttings taken early in the spring. Sow seeds indoors during any season. Maintain a temperature in the growing medium of 75°-80° until germination, which takes 25-30 days.

Fig. Firecracker Flower Crossandra infundibuliformis

FIRE LILY

Clivia miniata

The Fire Lily Plant, *Clivia miniata*, is native to South Africa.

Clivia miniata are sometimes known as the Fire Lilies because of their clusters of yellow throated, bright orange or red flowers that rise above the broad, 18" dark green, straplike leaves.

Unfortunately, Clivias are only hardy in USDA zones 10-11 and will not tolerate temperatures that drop much below 50°, so they are most commonly grown as house plants.

Growing a Fire Lily in your Garden

In the garden, they should be grown in a bright but lightly shaded area and kept reasonably moist during the growing season, but drier during dormancy. In their native environmment, Clivias reside in forests where their thick roots grow through the natural forest litter, rather than sending their roots down into the soil.

With that thought in mind, plant your Clivia in a shallow, spreading fashion and provide a thick blanket of compost as mulch.

Growing Clivias as House Plants

Clivia miniata is very well adapted to growing in containers, and will even bloom better and more consistently when they are pot bound than if they are growing in the ground. They should be planted in rich, well draining soil. Water thoroughly, then allow the soil to dry completely before watering again.

Never allow your Clivia to set in a saucer filled with water!

As a House Plant, keep your Clivia where it will receive bright, indirect light and never direct sun. When they are actively growing, Clivias need day time temperatures between 65° and 75°.

Clivia plants require a dormant period from October through January when they must be kept cooler (50°-55°) at night and only watered sparingly.

When new growth begins to emerge in early spring, increase the night temperatures to 65° and begin to water thoroughly again (...but let the soil dry between waterings).

Feed your Clivias monthly with a good house plant fertilizer.

Mist your Clivia occasionally during the growing season, and wipe the dust from the foliage with a damp cloth regularly.

Watch out for mealybugs, especially when growing Fire Lilies in the garden.

Propagating Clivia Plants and Growing them from Seed

The best time to repot or pot up Clivias is in the spring, when they are just starting to grow vigorously again. This is an excellent opportunity for you to liberate a 'Baby Clivia' from the clump, to share with a gardening friend. Try to disturb the roots as little as possible, use a sharp, clean knife cut and separate an offset

Plant it in a small clay pot but don't water for a few days to give the cuts to callous over. New roots will begin to develop within a couple of weeks. Clivias can be grown from seeds, however the seeds MUST be removed from the pods and planted while they are still fresh. Germination requires a soil temperature of 75° to 80° F. and will takes 30-40 days.

FLAMINGO FLOWER PLANTS

Anthurium species

Flamingo Flower plants are durable and fairly easy to grow house plants that will thrive for many years under ideal conditions and can even survive for a year or two in the most adverse situations.

They have large, deep green, lance or heart shaped leaves and produce long lasting, brightly colored, heart shaped flower bracts, resembling those of their close relative, the Calla Lily.

Flamingo Flowers begin blooming in early February, just in time for Valentines Day!

Growing Requirements for Flamingo Flower Plants

Anthurium plants are native to tropical regions of the Americas and are only hardy in USDA zones 11-12.

The other 99% of us will just have to be content with growing and enjoying them indoors.

As with most other house plants, you will want to replicate their native growing conditions as closely as possible when growing Flamingo Flowers indoors. These are jungle plants that live protected from the sun by a canopy of trees. In the home, Flamingo Flowers should be grown in a draft free spot where they will receive lots of bright, indirect light but NEVER full sun.

They prefer temperatures between 78°-85° during the day, with night temperatures about 10° cooler. Plant growth and flowering will cease at temperatures below 65°. Temperatures above 90° can cause the foliage to scorch and can even be fatal.

Ideally, the humidity should be 80% or higher. Insufficient humidity can cause your Flamingo Flower's foliage to lose its glossy sheen. Humidity can be increased by setting the planter on a tray filled with moistened pebbles or placing moist, fresh sphagnum moss around the plant.

A strategically placed aquarium will help raise the humidity as well.

Regular misting or sponging the leaves off with room temperature water is also very beneficial.

Flamingo Flower plants should be watered regularly and thoroughly, allowing the soil to only dry slightly before watering again.

Never allow the planter to set directly in water for more than a few minutes.

Over watering can quickly cause the leaves to turn yellow, which if not corrected immediately, will be followed shortly by root rot, which is almost always fatal.

Feed your Flamingo Flower yearly at the beginning of the growing season, using a slow release, granular fertilizer,

OR feed every month while actively growing, using a ¼ strength dilution a soluble, 3:1:2 ratio fertilizer. over-fertilize your Anthurium. This is one situation when less is best!

Remove spent flower bracts as soon as they begin to fade and turn brown, to encourage continued flowering.

Anthuriums are particularly susceptible to attacks from aphids, thrips and scale insects, as well as mealy bugs.

Inspect your indoor plants regularly to control these house plant pests as well as other common plant problems.

Flamingo Flower plants should be re-potted every 2-3 years.

When re-potting your Anthurium, use care to avoid damaging the fragile roots.

Choose a high quality, peat moss based potting mix, with a high perlite content to ensure proper drainage.

Propagating Flamingo Flowers and Growing them from Seed

Flamingo Flowers can be easily propagated by dividing the root structure when you repot your plant, making sure that each division contains at least two rhizomes.

Anthurium seeds need to be sown while they are still fresh and moist.

Sow seeds indoors in individual 4" pots, then cover the pot with a mason jar to retain high humidity. Maintain a constant temperature in the growing medium of 80° until germination is complete. Germination takes 20-30 days.

Flamingo Flowers

Fig. Anthurium andraeanum

Fig. Anthurium antioquiense

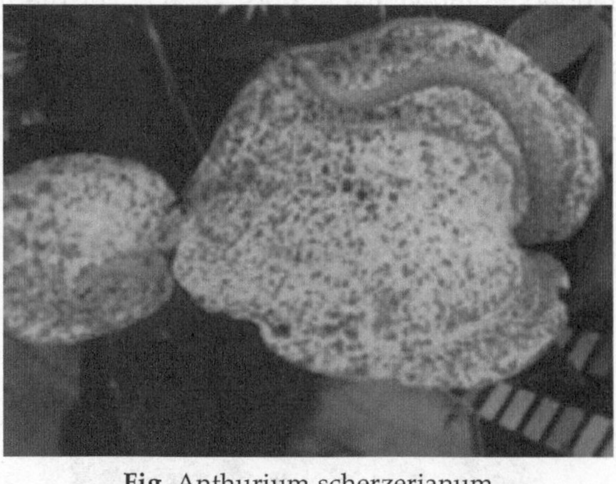

Fig. Anthurium scherzerianum

FLOWERING MAPLES

Abutilon hybridum

The Flowering Maple can be grown outdoors in warmer regions (Zones 8-10), but it is most often grown as a house plant.

The Flowering Maple is not really a Maple at all.

That common name is derived from their maple-shaped leaves.

Abutilon hybridum is also commonly known as a Chinese Bellflower or a Chinese Lantern.

The blooming season for most Abutilons is from April through June, but they will often flower periodically through the winter as well.

The flowers differ with each species, but generally resemble a cross between a Hibiscus and a Hollyhock, with colors ranging from white to reds, yellow, orange or several shades of blue.

The foliage is usually a pale green, but there are also variegated forms with white or yellow mottling.

Growing Requirements for Flowering Maples

In addition to needing filtered sun, Flowering Maples grow best when kept at about 70 degrees F. during the day and 60-65 degrees F. at night.

They require a well draining but evenly moist soil.

Flowering Maples benefit from being kept outdoors in filtered sun during the summer months, but must bereturned indoors before frost.

Be sure to check carefully for harmful insects, when you bring it back indoors!

During the winter, when growth has slowed, the soil should be allowed to dry out between waterings but never allowed to remain dry for too long.

Humidity is not too great of a concern except during the winter months when it is helpful to give your plant an occasional misting.

Feed your plant monthly when it begins showing signs of new growth, alternating between an all purpose *(20-20-20)* and a bloom type *(15-30-15)* liquid fertilizer applied at half strength.

Flowering Maple Plants tend to drop their lower leaves if they are underfed, so you will need to 'live and learn' to find the right frequency of feeding.

Flowering Maples should be repotted in the spring when it becomes necessary, but only into a *slightly larger pot* unless you want to drastically increase the size of your plant.

Growing Flowering Maples as House Plants

Flowering Maples need a considerable amount of room in which to grow.

If left unrestrained, they may grow to ten feet in height and spread, so it is important to not provide too large of a planter, nor to feed this plant too well.

When your plant has reached a suitable size you can control the growth somewhat by pinching, pruning and keeping it slightly root bound.

Flowering maples should be pruned to the desired shape in early spring or late fall. They can be trained into a hanging plant growth pattern by pinching back *each* stem tip every time is grows out two or three inches.

To create a more vertical tree like structure, allow only one main stem to remain, and remove all side shoots from it until it reaches the desired height. Then pinch the main tip to force lateral bushiness at the top of the plant.

Propagating Flowering Maples

Abutilons can be grown from seeds which will germinate in 21 days when kept at 75 degrees.

Seeds started indoors in early spring should be ready to produce their first flowers by the following winter.

Additional plants can also be cloned by softwood cuttings taken at any time of the year.

Cuttings will produce roots much sooner if bottom heat is available.

Fig. Flowering Maple Abutilon hybridum

GARDENIA PLANTS

Growing Requirements of Gardenia Plants

Gardenia requirements are very exact, and to insure your plants continued bloom, these needs *must* be met.

- Full sun, with some shading in the summer (in the hottest areas)
- An acid soil with a pH between 5 and 6
- Keep the soil moist, but never soggy
- Humidity, or daily misting
- Heavy feeding
- Cool night temperatures

Planting Your Gardenia

Gardenias should be planted in well conditioned soil containing peat moss and compost. Cultivating around the plants may damage their shallow roots, so the plant should be mulched with 2 to 3 inches of wood chips, sawdust or ground bark to keep the soil moist, cool and weed free.

Feed Gardenia plants every three weeks during the growing season with a Rhododendron & Azalea type food, or acidifying fertilizer such as Miracid®.

Proper temperatures are necessary for a gardenia to bloom.

This is the most critical aspect of getting your Gardenia to bloom. Flower buds will fail to form if day temperatures are higher than 70° or if night temperatures are over 65° or below 60°. The ideal temperature ranges would be 65 to 70° during the day and 60° to 62° at night.

Pruning your Gardenia

Gardenia plants can be pruned as far back as you like, but you should keep a few important details in mind.

First, cut plants back when they are dormant *(dependant on where you live)*. This will allow the stems to "heal their wounds" because the plants aren't actively growing.

This way you can still enjoy flowers the next year. When pruning, don't cut all the leaves off. Some leaves need to be left so the plant can still produce food for the root system. The following spring, use an acidic fertilizer with chealeted iron and high nitrogen so the plant can produce healthy leaves and buds. If you only want your plant to become bushier, simply pinch out the tips while the plant is dormant. Pinching will usually promote a heavier bloom the following year.

Growing a Gardenia Indoors

If you are growing your Gardenia as a potted plant indoors, keep it out of direct, hot sun, and allow it to only get bright light.

Black Mold on Foliage

Your 'black mold' is more than likely an algae growth, which is a cosmetic problem for your plant rather than a health problem. This could be caused by insufficient air circulation, or possibly lack of sun. Unfortunately, in your situation, the only way to remove this growth is by hand washing each leaf.

Fig. Lynne's Gardenias Gardenia jasminoides

GLOXINIA PLANTS

Sinningia hybridus

Gloxinias are fuzzy leaved plants that come in a wide range of flower colors, with variations in flower forms that include spotted and ruffled flowers and flowers with contrasting edges and throats.

Gloxinias are tuberous plants that are usually grown as house plants, but can also be grown in a shaded part of your garden.

Unfortunately, Gloxinias are only hardy in USDA planting zones 10-12.

In other regions you can grow your Gloxinia in a container and bring it back indooors before the first frost.

Your Gloxinia should be planted in moist, peaty soil and grown in partial shade or indirect bright light.

Over head watering should be avoided, because the velvety, trumpet shaped flowers tend to fill with water and collapse.

The foliage will die back after blooming at which time you can dig up the tuber and allow it to dry in a frost free location.

Growing Gloxinias as House Plants

Indoors, your Gloxinia should be grown where it will receive bright light but not direct sun.

They grow best in a humid area at 60°-75°.

Gloxinia plants are usually purchased already in bloom, but tubers are often available from December to March.

Plant the tuber (hollow side up) in a 4-5 inch pot, an inch deep in rich, light soil (a mix of loam, peat and sand works well) and water sparingly until the first leaves begin to grow.

Gradually increase watering as the roots fill the pot, using care to keep the water off the velvety foliage.

When the roots have filled the pot, replant it in a larger (6"-8") pot, and begin feeding it regularly with ½ strength, 'bloom' type fertilizer. The blooms will begin to appear in 3-4 months, and may last for as long as another 3-4 months.

When the bloom cycle ceases and the foliage growth stops, gradually decrease watering until the foliage has died off.

Store the tuber (still in its pot) in a cool dark place for 6-10 weeks, keeping the soil*slightly* moist. Repot the tuber after this dormant period into a larger pot with fresh soil.

Growing Gloxinia Plants from Seed

Gloxinias can be grown from seeds sown indoors at any time of the year.

The tiny seeds require light for germination, so do not cover them!

Sow them onto a bed consisting of a mixture of 1 part potting soil, 1 part peat and 1 part sand.

Keep the temperature in the growing medium at 70°.

Do not allow the soil to dry out at any time.

Germination takes 15-20 days.

Blooming takes 6 months.

Fig. Gloxinia Sinningia hybridus

GOLDEN SHRIMP PLANTS

Pachystachys lutea

Golden Shrimp Plants are also known as Golden Candle plants or Lemon Lollipop Flowers.

They are 18"-24" tall, shrubby plants that form clumps of upright stems that are each tipped with brightly colored, 4" golden yellow bracts that are adorned with small, creamy white flowers. The flowers are short lived, but the bracts will last for many weeks. They have deeply veined, lance shaped green leaves that grow to about 6" long.

In zones 10-12 where they are hardy, Golden Shrimp Plants are popular as landscape plants that bloom nearly non-stop all year. In cooler zones they are sometimes grown as annual plants, but are more often grown in pots or planters and kept as house plants.

Growing Requirements for Golden Shrimp Plants

In the garden, Golden Shrimp Plants should be planted in full sun to partial shade.

They should be grown in *slightly* acidic, fertile, well-drained soil.

Prune back hard *(by up to a half)* in early spring, to induce branching.

Golden Shrimp Plants are exotic and beautiful additions to any house plant collection.

Indoors, they will grow best when they are kept in a draft free area where they will receive full sun in the morning and bright, indirect light during the remainder of the day. They bloom best when daytime temperatures are 65°-80° during the growing season, but are happy with 55° during the winter months. Keep the soil evenly moist, using room temperature water that has been allowed to sit for 24 hours so that chlorine or other harmful chemicals have dissipated from it. Reduce watering in the winter.

Mist the foliage frequently during the summer months, using room temperature water. Feed once yearly, in early spring with a granular, slow-release fertilizer, or feed monthly from spring through fall with a balanced, soluble fertilizer. Golden Shrimp Plants should be repotted and pruned back yearly, in the spring. Use a good quality potting mix that has a high perlite content to ensure good aeration and drainage.

Pachystachys are prone to attacks from mealy bugs, as well as scale and spider mites, so be sure to inspect your Golden Shrimp Plant for these and other pests before returning it indoors after summering in the garden.

Propagating Golden Shrimp Plants

Golden Shrimp Plant can be easily propagated with 4"-6" softwood or semi-ripe cuttings taken from non-blooming shoots in late spring or early in the summer. Provide additional humidity for the cuttings by enclosing the pot with a plastic bag over a wire frame, or keeping the cuttings in a terrarium until they are well rooted and growing. Keep them in bright, indirect light or under fluorescent grow lights.

Remove dead leaves from the cuttings promptly!

Fig. Golden Shrimp Plant Pachystachys lutea

CROSS BEGONIAS

Begonia masoniana

Iron Cross Begonias have some of the most unique foliage in the plant world.

Each heavily textured, bristly, light green leaf bears a reddish-brown marking resembling the Iron Cross used on shields and flags during the Crusades of medieval times.

Their pinkish-white or pale green flowers appear periodically,

but they are fairly insignificant and usually go unnoticed.

Iron Cross Begonias are easy to grow house plants that enjoy spending the summer in a shady part of the garden.

Growing Requirements for Iron Cross Begonia Plants

Iron Cross Begonias are perennial, rhizomatous plants that grow from 8"-15" tall with 6" leaves.

They are only hardy in USDA zones 11 and 12, so in most cases, they are grown as house plants.

These Begonias grow best in bright indirect sunlight coming from a south, east or west window.

Fluorescent *grow lights* are also an excellent alternative to natural light for these plants.

When potting, set the plant with the crown slightly above the soil line. They should be planted in a porous, slightly acidic planting mix that contains some peat moss and leaf mold. African Violet potting mixes work well for Begonias.

Using room temperature water, keep the soil evenly moist *(never soggy)* when the plants are actively growing, but allow the soil to dry slightly between waterings when growth slows.

Keep water off of the foliage to help avoid powdery mildew and possible crown rot. Never allow your plant to stand in water.

Feed actively growing plants every two weeks with a fertilizer formulated for house plants, applied at the recommended rates.

Iron Cross Begonias thrive with daytime temperatures of 70° and 60°-65° at night. They will perish at temperatures below 55°

Provide additional humidity by setting the plants on shallow trays filled with moistened pebbles, especially during the winter months.

Begonia masoniana are susceptible to attacks from spider mites, mealy bugs and aphids.

Outdoors provide your Iron Cross with protection from the wind.

Keep them in partial shade. Do not let them dry out! Beware of slugs and snails!

Propagating Iron Cross Begonia Plants with Leaf Cuttings

The easiest way of propagating new Iron Cross Begonias is with leaf cuttings struck in a mixture of moist peat moss and perlite.

Water the soil well to be sure that it is settled and moist. Using a nail or small stick, make evenly spaced, slightly angled cutting holes at the same depth as the length of the stems.

Select a good healthy leaf and remove it (with stem) from the plant by reaching in as close to the crown of the plant as possible, grasping the stem and gently twisting it until it pulls free.

With a sharp, *sterile* knife, cut the stem at an slight angle *(try to match the angle of the leaf)*, 1"-1½" inchs from the leaf.

Insert your cutting into your rooting media deep enough that about ¼"-½" of the base of the leaf is covered by the soil mix.

Gently press the soil around the cutting.

Keep the rooting media moist, but not soggy.

Place the tray in a warm (70°-75°), brightly lit area or under a grow light.

The cuttings should begin to produce roots in about two weeks.

You should have a new plantlet in about eight weeks!

Fig. Iron Cross Begonia Begonia masoniana

JADE PLANTS

Crassula argentea

Fig. Jade Plants Crassula argentea

The Jade Plant originated in South Africa, and has been grown as a house plant in Europe and America for a hundred years. Generally, Jade plants are very easy and productive plants to grow, provided that their needs are understood and met.

Growing Requirements for Jade Tree Plants

Jade plants should be grown in very bright sunlight and low humidity.

Jade trees will sunburn if they are suddenly moved from subdued lighting to full sun, so if your plant is accustomed to dimmer light, be sure to move it into the sun in gradual stages.

Jade plants prefer an environment with temperatures between 55°F at night and 75°-80°F during the day. They will, however, tolerate temperatures down to 40°F.

Jade plants should be repotted every two to three years, using a quality commercial potting soil mixed 50/50 with coarse builders sand, with a scoop of bone meal added. The optimum soil Ph for growing Jade Plants is 6.5.

Watering and Feeding Your Jade Plant

Jades and all other members of the Crassula family are succulent plants, in that they have the ability to store water in their leaves, stems and roots.

Jade plants have an active and a dormant growing cycle. Watering and feeding should be determined by this growing cycle.

During the spring and summer months keep the soil *slightly* moist. Water liberally *(a good soaking)* once a week, then allow the potting mix to dry out completely before watering again.

Immediately remove any excess water from pot saucer. Feed your Jade plant with a 10-20-10 or 5-10-5 ratio soluble plant fertilizer every two weeks. *(African Violet fertilizer works very well for most succulents.)*

Keep your Jade Plant dry during the winter months while it is in dormancy.

Do not fertilize Jades from November through the end of March!

Getting your Jade Plant to Flower

Typically, all healthy, mature Jades will bloom, usually around Christmas in the northern hemisphere.

Their blooming is triggered by the natural shortening of the daylight hours.

If your Jade Tree is in a room that usually has lights turned on at night, it will more than likely fail to bloom for you. Sometime in September or early October, move your Jade plants, Christmas Cactusand last years Poinsettia into their 'no night light' environment. Find a suitable area that seldom has the lights on at night but receives good, natural light during the day.

JAPANESE ARALIAS

Fatsia japonica

Japanese Aralias are one of the easiest of all shade garden or house plants to grow and care for.

They have large *(up to 16")*, dark green, palmate leaves and form tropical looking, evergreen shrubs that can grow up to 6-8 feet tall and 6 feet wide, but are easily limited to a height of 2-4 feet with pruning.

Mature Fatsias produce rounded clusters of small white flowers on white stalks in the fall, which are followed by small, fleshy black fruit.

Fatsia Japonica plants are only hardy in USDA zones 8-11. Elsewhere they are grown in containers and moved indoors for the winter, or they are grown as year-round house plants. Gold and white variegated forms are available, but they are slightly more difficult to grow.

Growing Japanese Aralias in the Garden

In the garden, Japanese Aralias should be grown in a protected area with partial to full shade except in northern zones where they will appreciate a little morning sun. Hot summer sun will quickly burn the foliage. Fatsias should be planted in slightly acidic, moist but well-drained soil that has been enriched with a generous amount of compost at planting time.

Water regularly and thoroughly to keep the soil evenly moist during the summer, but watering should be reduced during their dormant, winter months.

Provide a heavy mulch during the winter in areas where their hardiness may be marginal.

Feed Fatsias monthly when they are actively growing, using a complete, balanced fertilizer.

Take precautions to protect these plants from slugs and snails.

Growing Japanese Aralias as House Plants

Fatsias seldom bloom when they are grown indoors.

When grown in planters, their size increases proportionate to the size of the pot.

Indoors, Japanese Aralias should be placed in a spot where they will receive bright, indirect light.

The ideal indoor growing temperatures are 65°-75° during the day and 50°-60° at night.

They should be planted in a rich, fast draining, commercial potting mix and kept evenly moist from spring through fall, but slightly drier in winter.

Never allow the soil to become soggy! Over-watering can quickly cause root rot.

Feed every 2 weeks while actively growing using a half strength dilution of a balanced, soluble fertilizer.

Fatsias can be kept small by cutting them back hard in the spring. Pinch back new growing tips to induce bushiness.

Wash the leaves and mist regularly to keep your plant happy.

Japanese Aralias require a dormant period in the winter, when they will benefit from cooler temperatures of 50°-55°. Allow the soil to dry out slightly before watering during this time.

Propagating Fatsia Plants and Growing them from Seed

Japanese Aralias can be propagated with stem cuttings taken from late spring up until mid-summer.

Sucker growths can easily be dug up and transplanted elsewhere.

In areas where they are hardy, Fatsia seeds can be sown directly into the garden in the spring or fall.

Sow seeds indoors at any time of the year. Maintain a temperature within the growing medium of 70° to 75° until germination, which will take 15-20 days.

Fig. Japanese Aralia Fatsia japonica

JERUSALEM CHERRY PLANTS

Solanum pseudocapsicum

Sharlene wrote and asked: *"Hi there! Received a Jerusalem Cherry - know it is poisonous - no animals or little ones near by.*

The Jerusalem Cherry belongs to a diverse family of plants that includes Nightshade, Cockroach Berries, Horsenettles, Tomatoes, Potatoes and Egg Plants.

Christmas Cherry plants, as they are also often called, are perennial ornamentals that are hardy in zones 8 and 9

Growing Requirements for Jerusalem Cherry Plants

Jerusalem Cherry Plants are most commonly grown as a potted house plants.

When grown as a house plant, Jerusalem Cherries require full sun to thrive, but they will *survive* in moderatly bright light.

Jerusalem Cherry plants are prone to drop their foliage and/or flowers if they are grown in a warm*(above 72°)*, dry environment so it is a very good idea to mist your Christmas Cherry often.

Jerusalem Cherries should be planted in a rich, well drained potting soil, and kept moist during their active growing cycle.

Feed with a liquid 5-10-5 'blooming houseplant' fertilizer every two weeks while the plant is growing vigorously.

Discontinue feeding as soon as your plant has finished blooming.

After all the 'cherries' have dropped, cut your plant back drastically, and next spring, after all danger of frost has passed, plant it outside in a partly sunny part of your garden.

If you are lucky enough to live in a frost free zone, your Jerusalem Cherry will develop into a nice 2-3 foot, woody shrub.

If you want to keep the plant going in frosty zones, you will need to dig your plant up each year, repot it, and begin the cycle all over again.

Fig. Jerusalem Cherry (Solanum pseudocapsicum)

FLAMING KATY KALANCHOE PLANTS

Kalanchoe blossfeldiana

Flaming Katies, or Kalanchoes as they are more often called, are easy to grow, winter blooming plants with waxy, green, scalloped edged leaves, and clusters of tiny, brightly colored flowers that continue for 6 weeks or longer. The true species *Kalanchoe blossfeldiana* grows up to 20" tall and has crimson red flowers but most Kalanchoes that are commonly available today are dwarf hybrids that only reach 8"-12" tall, with flowers in shades of red, pink, yellow and orange. Christmas Kalanchoes generally become available at nurseries early in the fall.

Growing Requirements for Kalanchoe Plants

Kalanchoes are only hardy in USDA zones 10-12, elsewhere they are grown as house plants or in a greenhouse. Kalanchoes are well adapted to growing indoors, preferring daytime temperatures of 70°-75° and nights at 60°-65°.They should be grown in a good quality potting soil and kept where they will receive bright, indirect light or sun. Kalanchoes are succulent plants that are capable of storing a considerable amount of moisture in their leaves, so they should only be watered when the soil is dry to the touch. Never over-water!

Kalanchoes dislike humid environments and tend to develop mildew if the air too moist for their liking.

Feed them monthly while actively growing, using a flowering house plant type fertilizer.

Remove spent flowers regularly, and your Kalanchoe will probably re-bloom again in 6-8 weeks. In the garden, your Flaming Katy will need protection from hot afternoon sun.

Bringing your Christmas Kalanchoe into Bloom

Like the Christmas Cactus and Poinsettia, the Christmas Kalanchoe will only set blooms when it is kept in complete darkness for a specific amount of time each day.

Once the day length decreases to approximately twelve hours, the plant recognizes that it is time to bloom. To get your Flaming Katy to bloom in time for Thanksgiving, keep it in a box or dark closet for 12-14 hours a day, beginning in early September.

For blooms at Christmas, keep your Kalanchoe in the dark beginning before October 15th.

Propagating Kalanchoe Plants and Growing them from Seed

Although they arc considered to be perennial plants, Kalanchoes tend to decline after their first year or two. Luckily, they are very easy to propagate and keep fresh, new plants growing.

Christmas Kalanchoes *may* form small plantlets along the margins of their leaves that can be easily removed and given their own pots.

Leaf cuttings and stem cuttings can be taken in late spring.

Kalanchoe seeds require light for germination, so DO NOT COVER them!

Start seeds indoors in by June 1st for blooming size plants at Thanksgiving, July 1st for Christmas blooms.

Maintain a temperature in the growing medium of 70°-75° during germination, which only takes 10 days.

Pinch young plants back lightly at 6-8 weeks.

Fig. Flaming Katy (Kalanchoe blossfeldiana)

LANTANAS

Lantana species

Lantanas are fast growing, shrubby, spreading or trailing evergreen plants that grow from 1-4 feet tall.

Trailing Lantanas, *Lantana montevidensis* can be used as ground covers and will spread as far as 6 feet and help control soil erosion.

Lantana plants produce flat-topped 2" clusters of bright, multi-colored tubular flowers in the summer and fall.

The flowers open pink, yellow, lavender or orange before turning to red, white, rosy-purple or orange.

It is not uncommon for several colors to be found in the same cluster.

Lantanas are easily grown in planters and are suitable for growing as house plants where they can be induced to bloom during the winter months by growing them in a sunny window and keeping the flower buds pinched off until late fall.

Growing Requirements for Lantana Plants in the Garden

Most Lantana Plants are only hardy in USDA zones 8-11, but some new hybrids are hardy in zone 7.

In areas where they are not hardy, they are grown as annuals or as House Plants.

Lantanas should be grown in a warm part of the garden where they will receive full sun.

Insufficient sunlight can cause powdery mildew

They should be planted in fairly rich, loamy, well-drained soil.

Water regularly and deeply during the first growing season to establish a deep, healthy root system.

After the first year, continue deep watering but water less often.

Over watering can cause flowering to decline.

Once established Lantanas are fairly drought tolerant plants.

Prune back hard in the spring to remove any dead wood and to improve the shape,

then *feed your Lantana lightly* before new growth begins, using a good all-purpose fertilizer.

Pinch the tips of new growth to induce bushier growth.

Growing a Lantana as a House Plant

When growing Lantanas as house plants, they should be planted in a rich, fast draining soil and grown in a bright, sunny spot with night temperatures at about 60°. Allow the soil to dry between waterings.

Take cuttings in the spring and keep the buds pinched off until fall to have Lantana flowers indoors during the winter months.

Check regularly for white flies and spider mites!

Propagating Lantanas

Lantana plants are fairly easy to propagate with stem cuttings taken during the summer. Dip the ends in rooting powder and push them into a small pot filled with sterile potting mix. Keep them in a warm, sunny place until the roots are well developed.

Start Lantana seeds indoors, 12 weeks before the last killing frost in your area if they are to be grown in the garden, or at any time for indoor use.

Maintain a temperature in the growing medium of 70°-75° until germination, which takes 6-8 weeks.

Fig. Lantana montevidensis

Fig. Lantana 'Irene'

Fig. Lantana 'Lemon Swirl'

LIPSTICK PLANTS

Aeschynanthus radicans

Lipstick Plants have thick, glossy, dark green leaves growing in pairs along their long, woody stems. Each vine is tipped with clusters of 2" tubular, scarlet red flowers emerging from maroon calyxes in mid-summer.

They are tropical, epiphytic vines that are native to Indonesia.

As house plants they are most often grown in hanging baskets but can be trained to climb up a three foot totem as well.

Growing Requirements for Lipstick Plants

Lipstick Vine plants are only hardy in USDA zones 11 and 12 where they should be grown in partial shade.

When they are grown as house plants, they are happiest when kept in lots of bright, indirect light or beneath fluorescent grow lights all year. Never keep them in direct sunlight.

The ideal growing temperatures are 70°-85° during the day and 65°-70° at night, but never below 65°.

Lipstick Plants bloom best when they are kept a bit rootbound.

They should be potted in a rich, well-drained, peat moss based planting mix and be kept evenly moist during the growing season, but allowed to dry out slightly between waterings in the winter.

They thrive in fairly high humidity, so they should be misted daily if the air is dry.

Lipstick Plants are susceptible to attacks from aphids, thrips, scale insects and mealybugs.

Regular pruning is necessary to keep your Lipstick Plant from getting leggy.

Prune the longest stems back to six inches above the soil *after blooming*, to induce new growth and branching.

New growth will sprout at the point of the cut.

Propagating Lipstick Plants and Growing them from Seed

Lipstick Plants can be propagated with 2" stem cuttings taken from new growth at the tips of non-flowering vines in the spring or early summer and rooted in a moistened mix of peat moss and perlite.

Layering is another, fairly quick method of propagating new Lipstick Vines. Pin a stem node to the soil in an adjacent pot or tray, using a piece of stiff wire bent into aU. Cover the pinned area with moist soil. Once the layer is well rooted, it can be severed from the parent and repotted.

Sow Aeschynanthus seeds indoors at any time, barely covering the very fine seeds. Maintain a temperature in the growing medium of 75° until germination, which takes 10-12 days.

Fig. Lipstick Plant Aeschynanthus radicans

MAIDENHAIR FERNS AS HOUSE PLANTS

Adiantum raddianum

Delta Maidenhair Ferns are elegant, fast growing, evergreen foliage plants with delicate, lacy fronds that are held on wiry, 6"-12" black stems. Maidenhair fronds are composed of bright green, triangular leaflets and can grow up to 12" wide and 18" long. The plants themselves will spread to about 24". They are long lived plants when they receive sufficient humidity and are not exposed to drafts, gas fumes or cigarette smoke.

Growing Requirements for Delta Maidenhair Ferns

Maidenhair Ferns are only hardy in USDA zones 10-12 so they are most often grown as house plants.

They grow best in a draft free area with bright, indirect light or filtered sun light and night time temperatures of 55°-60°.

In order to thrive, they must have high humidity and consistently moist soil.

Ideally, your Maidenhair Fern should be kept in a terrarium environment, but if that isn't practical you should provide additional humidity by setting the planter on a tray filled with moistened pebbles or placing moist, fresh sphagnum moss around the plant. A strategically placed aquarium will help raise the humidity as well.

Regularly misting the foliage with *room temperature water* is also very beneficial.

Keep the soil evenly moist, but cut back on watering slightly in the winter. The roots must never be allowed to dry out. Never allow potted plants to sit in water for more than an hour or two!

Feed monthly from spring through mid-summer, using a half strength solution of a soluble, house plant type fertilizer.

Remove dead or damaged fronds regularly by cutting the stem back to the base.

Maidenhair Ferns are occasionally attacked by scale insects and mealybugs, so check undersides of the fronds regularly.

When it becomes necessary, your Maidenhair Fern should be re-potted in the spring, using a rich, well-drained potting mix, in a planter with plenty of drainage holes.

Never use a pre-fertilized planting mix for container grown ferns!

This is a good time to propagate a few new plants.

How to Propagate Maidenhair Ferns

Maidenhair Ferns grow from rhizomes that spread under the surface of the soil, making them very easy to propagate by dividing the roots into three or four pieces in early spring. Each section should contain a *minimum* of two or three healthy fronds.

Ferns do not produce flowers, therefore they cannot produce seeds.

They reproduce with spores that form on the underside of the fronds.

Propagating ferns with spores is a fun and fairly easy method of growing new ferns, but it can take 3-4 years to grow a mature plant.

Propagating Ferns with Spores

Fig. Delta Maidenhair Ferns, Adiantum raddianum

NORFOLK ISLAND PINES

Araucaria heterophylla

Norfolk Island Pines are native to Norfolk Island

...That's somewhere between New Caledonia and Australia in the South Pacific.

They may reach a height of 80 feet in their natural habitat, *but they will maintain a reasonable size when grown in a container in your home.*

Growing Requirements for Norfolk Pines

The beauty and symmetry of the plant, as well as the soft flowing texture of the foliage will make you want to decorate it, and use it for your living Christmas tree!

Norfolk Island Pines may be grown in most temperature ranges which are suitable for people. However, they will do better when grown in a cool room with day time temperatures of about 65° f. and slightly cooler at night, but *never below 50 degrees!*.

Drafts, extremes in temperature, either hot or cold, and sudden changes in temperatures should be carefully avoided.

Norfolk Pines need bright light *but never full sun*. Preferably you should place your tree within four feet of a large window. Norfolks can *survive* with only household incandescent or fluorescent lighting if they receive a minimum of 16 hours of light each day. This may be effective in keeping the plant *alive* for a year or two, but its health will decline. There is no substitute for bright, natural light.

Turn your Norfolk Island Pine frequently to keep it symmetrical.

Norfolks should be watered thoroughly every week during the summer to keep the potting mix slightly moist (never soggy!) Water more often if your plant is drying out earlier. During the winter months allow the soil to dry *completely* before watering again. Preferably, you should use rainwater or water which has set for 24 hours to protect your tree from chemicals.

Feed monthly in the summer with a half strength solution of a soluble house plant fertilizer.

Maintaining your Norfolk Island Pine

Lack of sufficient moisture in the air will cause tip browning, needle drop, and eventually, the lower branches to die off. Misting is the best thing you could do for your Norfolk Pine, followed by any other method which would add humidity to the room like a humidifier or aquarium.

Norfolk Pines are one of the very few plants who enjoy being misted with COOLwater, rather than room temperature water.

Pruning and Potting Norfolk Pine Trees

Do Not Prune Norfolk Island Pines! The only pruning that should ever be done is snipping off brown tips and dead branches. Brown tips will never turn back to green, and unfortunately, once they have turned brown, growth will stop from that point.

Any other pruning you may decide to do will have the same effect... no new growth from that point. Norfolks resent replanting, so they should only be repotted every 3-4 years in the spring. Water thoroughly before starting the potting process. Use a fast-draining, sterile, commercial potting soil mix.

Norfolk Island Pines as Christmas Trees

If you should decide to decorate and use your Norfolk Pine as a living Christmas tree, be sure that the soil is kept sufficiently moist during that time when it is decorated.

Electric Lights have a severe drying effect on plants.

Be sure to mist your tree daily, or provide other sources of additional humidity.

Fig. Norfolk Island Pine (Araucaria heterophylla)

GROWING ORCHIDS AS HOUSE PLANTS

The Queen of Flowers

Fig. Lady Slipper Orchids

For many years most people considered orchids to be a 'greenhouse only' plant. Some varieties may be difficult to grow, and therefore best left to greenhouse culture, but there are many varieties which, when given the appropriate conditions and care will do quite well in a home environment. Unfortunately, due to the fact that there are over 1700 varieties of orchids, I will only be able to cover their basic, general care on this page. Contrary to popular belief, not all orchids require warm temperatures, in fact many do better with cool daytime temperatures of 60 to 70 degrees and 55 degrees at night. Others may take up to 80 degrees in the day and a drop to 65 or lower at night. Since orchids tend to be rather expensive compared to most house plants, make your choice of these beautiful plants wisely, and according to the conditions which you will be able to provide.

The basic common requirements of orchids

Most orchids which are suited for home culture are epiphytic jungle plants (in nature they grow perched high in the trees, clinging to branches or in the trunk apex of the tree). They derive their nutrients from the air, rain, and any decaying vegetation which the roots can contact. For success you will need to duplicate the native conditions as closely as possible.

Humidity

Orchids need high humidity of at least 50%. It is almost a necessity for the pots to be positioned above a tray filled with pebbles and water. The orchid pots themselves shouldn't set directly on the pebbles though, because excellent drainage is essential. It is best to invert a pot over the pebbles and then use that as a stand to set the orchid's pot on or hang the orchids above the tray. Misting the foliage is very beneficial, but this should only be done when there is sufficient light and air circulation so that the plants have a chance to dry before night.

Watering

Orchids should be watered about once a week when the soil becomes dry, but they must never be allowed to remain wet or soggy for more than a day or two at a time. It is best to use water which has been allowed to sit for 24 hours to allow the chemicals to dissipate, and always at room temperature. Feed your orchid with water soluble 'orchid fertilizer' every two weeks during the growing season.

Light

The light requirements for orchids vary with the individual plant. Four hours of sunlight is about the minimum required during the winter months if you want your plants to thrive and bloom for you. In the absence of natural sun you can provide the necessary light with the use of flourescent 'grow lights'. The 1200 foot-candles of light can be produced with four 48 inch grow lights suspended 6-12 inches above the plants. The lights should remain on for 14 hours each day.

Potting medium

I recommend using ready made orchid potting mixes which are available from most garden centers or from the orchid growers themselves. This is mainly because of the ease and availability, it is fairly inexpensive, and it is blended to the proper texture and acidity. If you desire to create your own mix, it can be a combination of any or all of the following components: osmunda fiber, ground tree fern stem, or ground bark.

Fig. Phalaenopsis Orchid

Fig. Onicidium papilio

Fig. Dendrobium Freesia

ORNAMENTAL PEPPER PLANT

Capsicum annuum

Ornamental Peppers are annual plants that are excellent for border plantings as well a for growing as a potted house plant.

Growing Requirements for Ornamental Pepper Plants

In the garden, Ornamental Peppers should be grown in full sun or light shade and planted 9-12 inches apart. If they are grown in pots, you can bring them indoors prior to the first frost, and continue enjoying the fruits through the Winter months. When Ornamental Peppers are grown as a house plant they should be grown in full sun. Keep their soil evenly moist.

Growing Ornamental Peppers from Seed

Sow the seeds of Ornamental Pepper plants into pots around the first of May if you would like your pepper plants to be fruiting at Christmas time.

Do not cover the seeds because they require light for germination.

Keep the rooting medium moist and at a temperature of 70°-75°

The seeds will germinate in about 3 weeks.

Once the plants reach about 8 inches, pinch out the plant tips and begin monthly feeding with an all-purpose fertilizer.

Fig. Ornamental Pepper Plants Capsicum annuum

PAINTED NETTLE PLANTS

Solenostemon scutellaroides

Fig. Painted Nettle (Solenostemon scutellarioides)

Coleus plants are durable and easy to grow.

They are best known for their bright colors, and variety of foliage forms.

Technically, they are a tender perennial but they are usually considered an annual plant by growers and seed producers. Coleus plants should be grown in the garden in bright, indirect light, or in partial shade. The lower growing dwarf varieties*(6-12 inch)* will create a colorful border, or you can use the taller *(up to 36")* types as a dramatic background planting.

Many Coleus plants will survive full sun exposure but the foliage color is often enhanced when they are grown in the shade.

Coleus are also quite striking when they are planted in a container and grown as a house plant. By removing the flower spikes as they develop, and keeping the plant pinched back, the Coleus can be kept growing as a *perennial* for several seasons.

These plants are a member of the *nettle family* but they are also a mint. They do not have the stinging properties of the common nettle. Coleus plants are sold in a variety of sizes ranging from bedding "six packs" to massive *specimen sized*hanging house plants. The small starter plants may seem bland and fairly colorless when compared to the larger specimens, leading you to grab for the big, fancy one. Even though you may think that your new thirty dollar plant has to be the most colorful and beautiful Coleus *ever*, never underestimate the potential of a Coleus seedling! Once the seedlings reach a certain point in their growth, they will develop additional coloring and foliage characteristics. The intensity of light which the plant receives will also have a direct bearing on the intensity of the foliage coloring. Some varieties may produce their best color in light shade, while others look best in bright lighting. Florescent "grow" lights seem to bring out even more intense and vibrant coloring. Unless you have an *immediate* need for a BIG house plant, you will be better off to choose a variety of the smaller ones, and watch them grow and develop.

Growing a Coleus as a House Plant

Coleus make nice house plants as long as they receive sufficient light and food.Your Coleus should be planted in a light, quick draining, commercial potting soil. Place it where it will receive several hours of bright light *(south window)* each day, or provide artificial "grow" lighting for best leaf color, and fullest plant.

Coleus plants will adapt to a wide range of temperatures above 55 degrees, but will grow best when they are kept between 70 and 85 degrees. Keep the soil evenly moist, but never soggy. Feed your Coleus plants monthly, with a diluted (50% mix) liquid house plant fertilizer. Flower buds must be pinched off as soon as they develop to prevent the plant from producing seeds. *(The flower spikes are very insignificant anyway....)* Once a Coleus is allowed to go to seed, it has completed it's life objective, and it will usually die.

Pinching may also be necessary to prevent leggy growth. Any time that the tip growth is removed, the plant's energy will be diverted to the lateral side growth, creating a much bushier plant. Coleus are very durable, so you can cut your plant back severely if needed *(almost back to the soil level)*.

Growing Coleus in the Garden

Coleus plants should not be set into the landscape until the minimum outdoor temperature is 50 degrees F. Although Coleus plants will usually survive in full sun, the foliage color tends to intensify in light shade when they are grown outdoors.

Plant them twelve inches apart in rich, moist, well-drained soil with a neutral or slightly alkaline pH. Feed monthly with a liquid all purpose *(10-10-10)* fertilizer. Pinch the center stems out when the plants are 4 to 6 inches tall to induce bushier growth, and be sure to pick off the flower spikes as they form.

If the soil is allowed to dry out, the foliage will wilt, but normally will recover quickly when additional water is provided. Water your plants thoroughly at planting time, and then mulch the entire bed to conserve moisture. The mulch will also help to heat up, and retain the heat in the soil, thereby helping your plants to get established in their new home.

Propagating new Coleus Plants

Growing a Coleus collection from seed is really quite easy and quick.

Seeds are inexpensive, readily available and can be sprouted and showing their first colors in as little as two weeks. Another advantage to growing your Coleus this way, is that each packet of seeds will contain many different colors of plants. Seeds from different producers may include different hybrid varieties as well.

If your intent is to grow your Coleus as a house plant, the seed can be sown indoors, at any time of the year. If the seedlings are destined for a garden location, they should be started indoors, at least 10 weeks before your last expected frost, so that the plants will be well developed when it is time to plant them outdoors. Coleus seeds are very small. They should be sown onto a layer of moistened, sterile potting soil in a shallow tray, and then covered with a thin layer of fine soil. The tray should be covered with a pane of glass or sheet of plastic to retain moisture, until the seeds have sprouted.

Keep the tray in a warm *(65-75 degree F.),* bright *(NOT full sun)* place. When the seedlings are large enough to handle easily, they should be thinned out and transplanted into individual pots. Seedlings should only be held by a leaf, never by the stem! When all danger of frost is past the plants may be set out in the garden.

Softwood cuttings from your Coleus

Propagating new Coleus with softwood cuttings is nearly as easy as growing them from seed You can create an identical clone of your favorite Coleus by taking softwood stem cuttings at any time of the year. If you take cuttings in the fall, prior to a killing frost, and grow your Coleus as a house plant to brighten the winter months. In February or March take several cuttings and cultivate them for transplant into the garden when the weather warms. Older Coleus plants tend to become leggy, and the foliage color will begin to fade. Propagation by cuttings will do a lot toward renewing most older plants, but the "cloned" baby will quickly become the better plant.

PEACE LILY PLANTS

Spathiphyllum floribundum

"I have a huge Peace lily, but for the past two years it has refused to flower, the tips of the leaves are going brown,there is some mold in the soil as well.

I have searched for answers to this but can never get the right one, can you help me.."

Growing Requirements for Peace Lily Plants

Unfortunately, Peace Lilies are only hardy in USDA zones 10-11. Most tropical house plants should be re-potted every 2-3 years, to freshen and revitalize their soil. The tip burn could be caused by over feeding, or chlorine in the water. Peace Lilies will grow and flourish in almost any well drained soil. However, it is always the best idea to use a good commercial soil mix containing peat moss, bark and sand.

Peace Lily Plants prefer bright filtered light, but will survive in low interior light. Peace lilies grow best in a warm environment with 68°-85° daytime temperatures and 10° cooler at night. Keep your plants out of drafts, and never expose your plant to extended periods below 40 degrees.

Feed your Peace Lily every 2-3 months, using a ½ strength solution of a balanced, soluble house plant fertilizer. Over fertilizing can burn the leaf tips and the roots. Peace Lilies are particularly susceptible to attacks from spider mites!

Watering Your Peace Lily

When watering, it is *very* important to use room temperature water which has been allowed to sit for 24 hours to allow the chlorine to evaporate. It is also a good idea to allow the water to absorb up from the bottom to keep water off of the foliage. The soil should be kept evenly moist but never soggy. Allow the soil to dry out *slightly* between waterings, but excessive drying can make the plant wilt and cause yellowing of the leaves. Cleaning the leaves of your Peace Lily regularly with a damp cloth to remove dust will not only help the appearance but will also remove some pests which tend to gather on the underside of the leaves and help the plant to breathe.

Fig. Peace Lily Plants (Spathiphyllum floribundum)

PERUVIAN LILY PLANTS

Alstroemeria

Peruvian Lily plants are tuberous rooted perennials with bright green, lance shaped foliage that grow 2-3 feet tall. They produce loose clusters of speckled, 1½"-2" Azalea-like flowers on long, leafy stems in late spring into mid summer.

Alstroemerias are low maintenance, long lived plants that are prized for their usability in cut flower bouquets. There are about fifty species of Alstroemeria that were discovered in South America in the 18th century and brought back to Europe where botanists quickly began to experiment and create new hybrid cultivars. Most commonly available Peruvian Lilies are hybrids of *Alstroemeria aurea*, *A. ligtu* or *A. chiliensis*.

Growing Requirements for Alstroemeria Plants

Peruvian Lily plants are hardy in USDA zones 7-10. They should be grown in full sun except in hot summer regions, where they will do fine in light shade. Peruvian Lilies should be grown in rich, light, slightly acidic soil. Prepare the planting hole by adding generous amounts of compost and peat moss to the soil along with coarse builders sand for added drainage. Alstroemeria tubers should be planted 6"-8" deep, using care not to damage their brittle roots.

Water regularly and thoroughly to keep the soil evenly moist. In cold winter regions, apply a heavy mulch of peat moss or dry leaves in the fall. Top-dress with well rotted manure or compost as soon as new growth begins each spring. In areas where they aren't hardy, the tubers should be dug up in late fall and stored in some damp potting soil and kept in a cool, dark area. Alstroemerias are suitable for growing in pots and can be maintained as house plants if you can provide them with a cool space and sufficient sun. Water regularly, but be careful not to over-water. Reduce watering in the winter months. Feed every 2-3 weeks when in bloom, using a ½ strength solution of a soluble, bloom type fertilizer.

Pick Alstroemeria Flowers, Do Not Cut Them

Cutting Alstroemeria flowers will slow the growth of the plant. The proper method of harvesting Peruvian Lily flowers is to grasp the stem a few inches from the ground and twist it as you pull it upward. This will detach the flower cleanly from the rhizome and promote heavier flower production.

Propagating Peruvian Lilies and Growing them from Seed

Alstroemerias should be left undisturbed for as long as possible before digging and dividing them because they will take 2 or 3 seasons to re-establish themselves after transplanting.

They should only be divided in early spring. Alstroemeria seeds can be planted directly into the garden in early spring, once the temperature reaches about 40°. In mild wither regions, they should be planted in the fall. Sow seeds indoors in late winter.

Maintain a soil temperature of 55° until germination, which takes 15-20 days. Once the seedlings are 5"-6" tall, give them a light feeding of a slow release fertilizer.

Fig. Peruvian Lilies (Alstroemeria)

PIN CUSHION PLANTS

Nertera granadensis

The Pin Cushion Plant is also commonly known as a Bead Plant or Coral Moss. It is a fast growing tender perennial that produces an abundance of tiny orange berries just in time for Halloween!

Growing Requirements for Pin Cushion Plants

Pin Cushion Plants are only hardy in zones 9-11, however, they grow best in cooler conditions... Pin Cushions require full sun or they will produce a lot of foliage and very few berries.

In hot summer regions, they maay need partial shade.. They should be planted in a sandy, well draining soil. Give your Pin Cushion plant a half strength dose of a good all purpose fertilizer in early spring and mid summer. Keep the soil evenly moist but not soggy.

Growing a Pin Cushion Plant Indoors

Pin Cushion Plants are an excellent choice for growing in containers. If they are grown as house plants Pin Cushions enjoy spending their summer months outside in a sunny location, but must be moved back indoors before the temperature drops below 55 degrees F.

It should be kept in a sunny location with a room temperature of 65-70 degrees. When active growth slows during the winter, watering should be cut back. Never allow the pot to stand in water!

Propagating and Growing Pin Cushion Plants from Seed

Sow Bead Plant seeds indoors at any time if you intend to grow them in a greenhouse or as house plants. Nertera seeds require darkness for germination so be sure to cover them well! Keep the temperature within the growing medium at 65°-70°

Germination will take place within 20-25 days. Pin Cushion Plants can also be easily propagated by division.

Fig. Pin Cushion Plants (Nertera granadensis)

PITCHER PLANTS

Sarracenia

There are eight species of Sarracenia, The North American Pitcher Plants. These prehistoric looking plants are characterized by leaves that curve into tubular funnels which catch and digest insects. Pitcher Plants range in height from 4" long *(Parrot Pitcher Plant)* to more than 3' tall*(Yellow Pitcher Plant)*.

Pitcher plants attract their prey with an intoxicating nectar which is secreted from the hood of the funnel. Once the insects enter the pitcher, escape is nearly impossible because of the downward pointing hairs on the inside of the hood and the slippery wax coating on the inside of the Pitcher's tube *(not to mention the fact that they may be a bit drunk)*.

The unfortunate insect meets it's doom when it falls into the pool of liquid in the base of the funnel, and is digested by the enzymes which the plant excretes. The beautiful coloring of the North American Pitcher Plant will be much more intense if the plant receives at least a couple hours of bright, full sunlight each day, but will also grow reasonably well in partial shade. A temperature range of 60 to 85 degrees F. is ideal. Pitcher Plants should be grown in a mixture of live Sphagnum moss, medium orchid bark and charcoal. If you are unable to get any living Sphagnum moss, a mixture of peat moss and coarse builders sand will suffice.

Make sure the soil is kept very wet during the growing season, from May through October. *Only* rain water or distilled water should be used for watering carnivorous plants, and they should NEVER be fertilized...

Pitcher Plants require a dormant cycle in the Winter. When active growth stops in late fall, clip off all of the dead or dying pitchers, leaving the smaller (still developing) pitchers. Keep them in an area that stays about 40 - 45 degrees for about 3 months, until new growth begins in the Spring.

Pitcher Plants can be divided and repotted as the plant is coming out of dormancy, but this must be done before vigorous new growth begins. Most Pitcher Plant species are hardy in USDA zones 6-8, and may survive in even colder areas provided they are given a thick cover of winter mulch. They make excellent and interesting additions to any bog garden. Before you leave please read How to Grow Carnivorous Insectivores.

Fig. North American Pitcher Plant Sarracenia purpurea

PRAYER PLANT

Maranta leucoreura

One possible cause of the burned leaf tips on your Prayer Plant could be chlorine or flouride in your water supply. This problem can be alleviated by using bottled water, or allowing the water to sit in an open container for 24 hours before using it.

Growing Requirements for a Prayer Plant

Unfortunately, the Prayer Plant is only hardy in USDA zones 11-12. Prayer plants need very high humidity to thrive. They grow best in a warm greenhouse or a terrarium. Provide medium or bright filtered light, but never direct sun. Plant your Prayer Plant in a well drained soil to which a little peat moss has been added, because they need a slightly acid soil. Keep the soil evenly moist but never soggy except when Prayer plants are resting in winter, when you should keep the soil drier. Feed your Prayer Plant a balanced acid type fertilizer like *Miracid*® every 2 weeks when actively growing.

Prayer Plant Growing Tips

Repot your Prayer Plant in early spring if needed, at which time you can propagate new plants by division. The ideal temperature range for growing Prayer Plants is between 60° and 85°.

Dry air may be a problem in the winter, and this is often the cause of brown leaf tips. Putting the Prayer Plant in among several other house plants will help to create a more humid atmosphere. Another aid for humidity would be to place a bowl of water in the immediate vicinity of the plant or place the pot on top of a saucer filled with moist pebbles.

Fig. Prayer Plant (Maranta leucoreura)

PURPLE PASSION PLANTS

Gynura aurantiaca

The Purple Passion or Purple Velvet Vine is a trailing, evergreen perennial that was originally discovered on the island of Java in Indonesia.

Growing Requirements for Purple Passion Plants

Unfortunately, Purple Passion Plants are only hardy in USDA zones 10-12, so their most common use is in hanging planter, growing as house plants. In the garden, plant your Passion Vine in an area where it will receive some shade from the hottest, mid-day sun. It should be grown in a rich, slightly acidic, well-drained soil.Keep the soil evenly moist during the growing season. The vines are capable of growing up to 8 feet in length but will lose much of their bright purple coloring. To retain the rich purple coloring, keep them trimmed to a length of 3 ft. *or less*. Pinching the tips will result in a much fuller plant. Fertilize monthly during the growing season with a balanced fertilizer.

Growing a Purple Velvet Plant as a House Plant

Purple Passion Plants need bright light for the brightest coloring but they must be protected from hot afternoon sun. They prefer a slightly acidic soil mixture, so you should use a soil mix consisting of 2 parts peat moss to 1 part potting mix and 1 part coarse sand. The soil should be kept evenly moist at all times during the growing season. Feed every 2-3 weeks with half-strength house plant food when the plant is actively growing.

Pinching the tips regularly will help produce a fuller plant. Purple Passion Plants produce yellowish-orange, 1" flowers in mid-summer. These flowers have an unpleasant odor, so you may want to remove them in their bud stage. During the dormant winter period, light and watering can be reduced slightly. Leggy vines should be cut back to 3-4 nodes to produce a much fuller plant the following spring.

Propagating Purple Passion Plants

Purple Passion Plants can be easily propagated by taking softwood cuttingscontaining two stem nodes each, from actively growing vines. Plant several cuttings directly into a six or eight inch pot, filled with the potting mix mentioned above. Place the potted cuttings in a warm, brightly lit area.

Fig. Purple Passion Vine Gynura aurantiaca

SPIDER PLANTS

Growing Requirements for Spider Plants

Spider plants need bright natural light but they will sunburn if grown in direct sunlight. Spider Plants also thrive when grown under artificial grow lights, but they will be reluctant to produce the plantlets unless you decrease their light hours to simulate the naturally shorter days of fall, when they normally produce their 'babies'. Keep the soil evenly moist, but never soggy. They prefer to be grown in a cooler (55°-65° F.) room but will tolerate warmer temperatures. Feed your plant every 3 or 4 months with any house plant fertilizer. Damaged

Spider Plant foliage can be trimmed with a pair of clean, sharp scissors if necessary.

Maintaining Your Spider Plant

Spider plants are generally easy to grow, and mature plants are quite beautiful... until the tips brown out!

Tip burn is usually caused chemicals in the water.

If you are using city water for your plants it has been chlorinated, and very likely has been treated with fluoride as well, either of which chemical will cause the tip burn.

Using rainwater or distilled water will take care of this problem.

Tap water should be allowed to set for at least 24 hours before using it to water a Spider Plant.

Over fertilizing can also seriously damage your plant, so when you feed it, use ½ of the recommended strength.

Propagating Spider Plants

Spider plants are very easy to propagate.

The ideal method is to allow the plantlet to root while it is still attached to the mother plant.

Set a pot filled with sterile potting soil within reach of one of the plantlets and just place the offset onto the surface of the soil.

If necessary you can pin it in place with a bent paper clip.

When the young plant has rooted in 7-10 days just sever its ties with the mother.

If it is impossible to use this method, look at the base of the plantlets and choose one with a fleshy, healthy set of roots rather than choosing one with good foliage.

Sever it from the parent and follow the same procedure, pinning it to a pot of soil. This method of rooting will take about a month.

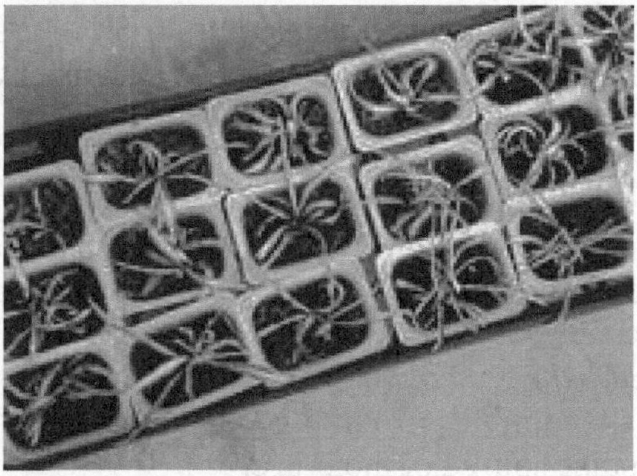

Fig. Spider Plants (Chlorophytum comosum)

VENUS FLY TRAPS

Dionaea muscipula

The Venus Fly Trap *(Dionaea muscipula)* is the most well known species of carnivorous plants.These insectivorous plants lure their prey using a sweet smelling nectar. When an insect lands on the head of the fly trap, they seek the source of the nectar, and if the plant is lucky, it will touch one of the many trigger hairs located within the jaws of the trap. Once triggered, the trap snaps closed, trapping the victim. The plant then secretes an enzyme which essentially 'dissolves' the insect, turning it into a digestible dinner. After a few days, the fly trap will re-open, to await it's next victim.

Growing Requirements of Venus Fly Trap Plants

Venus Fly Traps are quite easy to grow, as long as they are given the proper growing conditions. Flytrap plants should be grown in very bright light, but not in direct, hot sun. A warm, humid environment, with a constant supply of moisture, like you would find in a terrarium is ideal. The potting mixture should consist of a mix of 70% peat *or sphagnum moss* and 30% perlite or coarse pumice. Venus FlyTraps are extremely sensitive to chemicals, so you should only use distilled water or rain water. If it is absolutely necessary to use tap water, allow it to set for 24-48 hours in an open container, to allow the chlorine and any other chemicals to dissipate.

Fly Trap plants do not need fertilizer, the insects which they consume provide them with all the nutrients they need. Fly Traps Require a period of Dormancy Venus Fly traps have an active growing and feeding season, from May through October. During this period the soil should be kept constantly damp. Temperatures should be kept between 70° and 95° F. During the winter, Fly Traps require a dormant period of about three months. At this time, much *(if not all),* of the foliage will die back. The entire planter should be moved to an area where the temperature will remain at 45-50 degrees F., or the bulb may be removed, sprayed with fungicide, wrapped in damp, live sphagnum moss, placed in a plastic bag, and moved to a cool area.

Venus Fly Trap Growing Tips

If your plant is *not* grown in a terrarium, it is a good idea to place a pie pan or large saucer, with about an inch of water in it, under the pot. *(Elevate the pot by placing pebbles under it so that the base of the pot is barely in contact with the water, not submerged... the growing medium must stay moist, but never soggy!)*

This will keep the humidity around the plant higher and it will ensure that the plant has a constant source of moisture.

- If your plant is growing in a terrarium, you will have to provide it's food.
- Your Fly Trap will consume 2 or 3 small insects each month. Dead flies and insects can be used, provided that they died of natural causes and not by poisons.
- Never, *never*, never feed your Venus Fly Trap any hamburger. The fat content in burger will be fatal to your plant..
- Artificially springing the trap, *i.e. poking it with your finger,* drains the FlyTrap's energy. If this is done too often, the fly trap head will become less sensitive and *possibly* die.
- Always keep dead leaves and heads cut off to prevent fungal infections.

Fig. Venus Fly Trap (Dionaea muscipula)

6

Indoor Gardening and Plant Care

INTRODUCTION

Summer is over and the garden outside is finished for the year. All the plants outside have either gone dormant for the winter or finished their life cycle. But gardeners who still crave the satisfaction and rewards of growing things can put their gardening talents to work indoors during the winter.

Whether your gardening need itches during the winter or not, having plants in the house can add to the overall health of your household. Having one plant for every 100 square feet in your house helps remove indoor pollutants. In addition, plants take up carbon dioxide and give off oxygen as a waste product.

Humans, of course, take up oxygen and give off carbon dioxide as a waste product. So, having plants in the house makes for a healthy and equitable exchange for humans and plants.

A fancy greenhouse is not necessary to grow vegetables and beautiful flowers indoors. Many plants can thrive in your house and there are several techniques to help you keep your green thumb working all winter long.

LIGHTING

One of the first things you will need to address for your indoor gardening is lighting. No matter what else you do for your plants, if they do not have the proper lighting they will not thrive. Lighting is, in fact, the most important aspect of indoor gardening. When choosing decorative plants for your home, look for plants that require medium to low light, like Philodendrons, some ivy and some ferns.

Plants And Indoor Light

Plants that are not getting enough sunlight will generally have thin stems and smaller leaves. The leaves will also be lighter on a plant that is not getting adequate light. Most plants with medium light requirements will benefit from a south facing window. Windows that face east and west will generally give plants with medium light requirements adequate sunlight, as well. However, plants that prefer low light will probably prefer a north facing window. Because the sunlight comes through your windows from one side only, even when placing your plants near windows with the proper amount of sunlight you will still need to make sure that all sides of the plant get the full benefits of the light.

The foliage on the side of the plant that is facing away from the window will reach for the light, growing up and over toward the window and making your plant look uneven. The foliage on the side of the plant facing away from the window may even be a lighter color. Each week, rotate your plant so that a different side of the plant is facing the window. This will allow all the foliage on the plant to grow evenly and healthy.

If the windows in your home do not let much light into the house, if they are shaded by trees or buildings or if the angle from the sun just does not allow much light into the room, you can use artificial light for your plants. Most hardware stores and some supermarkets carry incandescent grow lights that will screw into a regular socket and will provide your plants with supplemental light when there is some, but not enough, sunlight. There are also a variety of specialty grow lights that will also provide the necessary light for growing vegetables inside and for starting spring seedlings inside.

Specialty Grow Lights

- One type of specialty grow light is an HID (High Intensity Discharge) light. HID lights come in two types; the Metal Halide (MH) bulb, which produces lighting in the blue spectrum and is ideal for foliage growth and, High Pressure Sodium (HPS) bulbs, which give a red-orange light, ideal for increasing bud and flower production. The MH bulbs can be used as a primary or single lighting source while the HPS bulbs are best used as a secondary or additional lighting source. MH and HPS bulbs can be used together to give your indoor plants the full spectrum of light.
- Compact fluorescent full-spectrum lights are now available that, especially when used in multiples, can provide some plants, such as African violets and herbs, with all the light they need.

Most hardware stores will carry a wide variety of specialty grow lights and they can also be found online by searching for grow lights.

Reflectors

Reflectors can be an effective way of maximizing the use of electric grow lights. Place reflective Mylar or white paper around the edge of the containers to reflect the light back up onto the plant. Aluminium foil will also reflect heat back up onto the plants and should not be used as a reflector. Using an electric timer on your grow lights will keep your plant's day and night cycles regulated. Plants do need a night cycle for rest in order to grow properly. When using banks of artificial lighting for your indoor plants, care must be taken to provide adequate ventilation. The heat from multiple electric bulbs can build up and be detrimental to the plants. Leave a window cracked for several hours a day, if possible, or run a ceiling fan or portable fan to circulate the air.

WATERING

Proper watering of your indoor plants is as important as proper lighting. More indoor plants are killed by over-watering than by any other cause, so it is important to know the water requirements of each plant that you have indoors. It is also important to note that plants that are grown indoors in containers will require more frequent watering than plants grown outside in the ground because the containers restrict the growth of the root system. Electric lighting also creates a drying effect that creates a need for more frequent watering.

TYPES OF WATER

Indoor plants can be watered with regular tap water. The chlorine and fluoride that is added to most city water will not harm your house plants. However, do not water your plants with water that has been run through a water softener, especially a water softener that uses salt. If you do buy bottled water for your plants, buy regular spring water or drinking water, not distilled water. Distilled water has had all the minerals removed and your plants do need the minerals that are naturally found in drinking water. Use water that is the same temperature as the room the plant is in. Using water that is too cold or too hot can shock your plant's root system.

The humidity in the room is also an important factor to consider in your plant's water needs. If the room air is very dry, consider using a humidifier. Air that is too dry for too long can cause the leaves of the plant to dry and curl at the edges. You can also mist the plants leaves, depending on the plant, if the air in the room is too dry. Mist the leaves early in the day to prevent them from rotting overnight.

When To Water

Not all plants have the same water requirements. Tropical plants may require more water than Mediterranean plants and flowering plants will require more water than plants that do not flower. Carefully read the watering instructions for all the plants or seeds that you purchase. Along these same lines, not all of your plants will need watering at the same time. The size of each plant, the amount of light and ventilation each plant gets, the temperature and humidity of the room all have an effect on how much water a plant needs.

Watering all of your indoor plants at the same time each week is not a good practice. Instead, learn to tell when your plants need water by the feel and color of the soil. As the soil dries out it will become a lighter color and will eventually begin to pull away from the sides of the container and even crack. A good way to check the moisture content of the soil is to stick your finger into the soil one or two inches deep. If the soil at the bottom of the hole is dry, your plant needs water. Do not water your plants until the soil is dry enough that the plant actually needs water. When the soil is too wet the plant can not get enough oxygen and will literally drown. The roots will turn brown and will rot or will be subject to disease and the leaves will turn yellow and will fall off.

While it is better to let a plant get a little too dry than to over–water it, letting a plant stay too dry can also kill the roots. If you are away from home often or find it difficult to gauge when your plants need water, you can invest in self-watering planters that will constantly give your plants the proper amount of water. Do not water your indoor plants at night. Transpiration, the process where water vapour passes through the leaves of the plant, is an important part of the watering process and can not occur at night in indoor plants. Do not automatically assume that the plant needs watering if it appears stressed. Carefully check the soil's moisture by sticking your finger as far down into the dirt as possible before you water a plant that appears stressed.

How To Water

There are two methods you can use to water your plants. One method is to water your plants from the top of the soil. When watering from the top of the soil, use a watering can with a narrow spout that will fit between the leaves to prevent splashing too much water on the foliage. Water all over the top of the soil, not just in the middle on or the sides.

You can also water your plants from the bottom. For some plants that do not tolerate moisture on their leaves, such as African violets, this is the best method. This method is also preferred for most plants because it allows the water to soak up through the entire soil base. Many plant containers have large saucers that go underneath the pot. Fill this saucer with water and allow the soil to soak up the water through the drainage holes. Allow the water to stand for one hour and then drain off any remaining water. Do not allow the remaining water to stay in the saucer. This will result in over-watering and prevent the roots from receiving the oxygen they need. If you container does not have a saucer, fill a sink or large bucket partially with water and place the plant container in the sink or bucket. Make sure that the water does not flow over the top of the container. Let the plant sit in the water until the top of the soil is moist. Plants that are consistently watered from the bottom may eventually show signs of salt build-up on the top of the soil. If this salt build-up is moderate, water the plant from the top of the soil thoroughly and discard the water that drains out of the soil. If the salt build-up is heavy, scrape off the top layer of the soil, taking care not to disturb the roots, and replace it with fresh soil.

FERTILIZING

Most indoor plants will not require frequent fertilizing. A good application of fertilizer every two to three months should be sufficient for most indoor plants.

Do not fertilize new plants for at least two months because they will have already been fertilized at the nursery. Carefully read the printed instructions that come with your plant or research on your own to make sure you know the fertilizing needs of each particular plant.

There are many good commercial fertilizers on the market made specifically for house plants. Carefully follow the label directions when fertilizing your plants and you will avoid over and under feeding your indoor plants.

Plants that have been over fertilized will have dried or burned leaves. If you feel you have over fertilized your plants, water them thoroughly from the top and discard the drained water for two to three weeks.

ROOM TEMPERATURE

Not all plants have the same temperature requirements. Carefully read the instructions that come with your plant for any specific temperature requirements. As a general rule, most house plants do well in day time temperatures of 65° F to 75° F and 50° F to 60° F night time temperatures.

These temperatures can vary as much as plus or minus 10 degrees for most plants. Plants with different temperature requirements will generally have that noted prominently in the printed planting instructions that come with the plant. Major temperature fluctuations are harmful to indoor plants. Try to keep the temperature as stable as possible in the rooms where your indoor plants live. The leaves will turn yellow and fall off on plants that are grown in temperatures that are too cold. A plant that is grown in temperatures that are too hot will be small and will have weak stems and leaves.

CHOOSING THE RIGHT CONTAINER

SIZE

It is important to choose a container that will give your indoor plants plenty of room to grow and small enough that the plant will not put all its energy into expanding its root system to fill the available space. For example, when transplanting a four inch potted plant into a new pot, the new pot should be no more than two to four inches larger.

When transplanting a twelve inch potted plant, the new pot should be no more than four to six inches larger. When you grow your indoor plants from very young plants, you may need to transplant three or four times before the plant reaches its mature size.

DRAINAGE

All plant containers should contain drainage holes in the bottom of the pot. Plants that are not allowed proper drainage will suffer from oxygen deprivation and the diseases that can effect roots when they are kept too wet. Clay and pottery pots are excellent containers for indoor plants because the pots absorb some moisture; however plastic containers also work very well. Whatever material the container is made from, it must have proper drainage holes in the bottom of the pot.

If the container you have chosen does not have drainage holes, drill or cut several holes about the size of a pencil into the bottom of the pot to allow excess water to drain. Space the drainage holes evenly all around the bottom of the pot to allow for even drainage. It is best to purchase a large saucer with your plant pot. This will allow you to water your plants from the bottom, if you choose, and will also keep freshly watered plants from dripping on to furniture and floors.

POTTING AND REPOTTING

The first step in repotting indoor plants is to make sure the new containers are clean and free of insects and disease. Even if the containers have never been used before, it is a good practice to wash them well with regular dishwashing liquid and a couple of teaspoons of bleach. Rinse the pots thoroughly with clear water until all traces of the bleach smell are gone.

Use a good commercial potting soil. Garden soil is not suitable for indoor plants although you can mix in up to one-half pure compost with the potting soil. There are also commercial potting soils now that contain timed release fertilizers which are excellent for use in repotting indoor plants.

The soil should be dark and moist with no clumps or rocks. If necessary, put the soil through a sieve to break up hard lumps and sift out any rocks or debris. If the soil is dry, place the soil is a bucket or pan and water it lightly, mixing the soil around in the bucket or pan to get it all moistened. Let this moistened soil stand for an hour or two to make sure the moisture is evenly distributed. If you add too much water, just let the soil dry for a day or two before potting your plants.

TRANSPLANTING

With the moist, sifted soil, fill the pot about one-third with soil. Grasp the plant by its stem and gently slide it out of its existing pot, dirt and all. Then place the plant, with the dirt, into the new pot, pressing it down lightly. The plant's existing dirt and root ball should come up to about one inch below the rim of the pot. This will allow you to water from the top of the pot occasionally or consistently, if you desire. The plant's stem should not be covered with dirt or it will rot and the plant will die.

If the plant sits too low in the pot, take it out and add more dirt to the bottom of the pot until the plant's root ball and existing dirt sits about one inch from the rim of the pot. Carefully fill in soil around the plant to about one inch below the rim of the pot, making sure to keep the existing dirt and root ball level with the new dirt in the pot. Press the soil firmly but gently down as you fill the pot to prevent air bubbles and excess settling.

Water carefully from the top, making sure that the water doesn't sit on top of the soil. When your indoor plant has become root bound it will stop growing. Grasp the plant by the stem and slide it from the pot, dirt and all. It the roots are tightly packed and covering the outside of the dirt and growing out through the drainage holes, the plant is root bound and the roots will need trimmed before you repot it.

With a sharp pair of scissors or pruners, cut away about half of the tightly packed roots. Then loosen the remaining roots on the outside of the dirt so that they will grow down into the soil of the new pot and repot normally.

Planting Cuttings

When you have taken cuttings from other plants and rooted them in water there comes a time when they need to be transferred from the water to soil. Fill the new container about one third full of the potting soil and place the cutting inside the pot, holding it so that the bottom of the roots drag slightly onto the soil. If the stem of the cutting does not come to about one inch from the rim of the pot, add more soil until it does. Then carefully fill the pot with soil, holding the cutting straight and lightly packing the soil around it up to about one inch from the top of the pot. Water carefully from the top, making sure that the water does not stand on top of the soil.

GROWING AN INDOOR HERB GARDEN

Growing herbs indoors is simple and enjoyable. Simply select a spot in your home with ample sunlight or prepare artificial lights to either supplement or fully provide the light your herbs will need. Compact fluorescent lights can supply all the light your herbs need if used within two feet of the plants. Depending on the number of plants you plan to grow, you may need more than one light. Choose a container that will provide ample room for the mature plants and that will give your herbs proper drainage. A 16 inch rectangular pot will have enough room for about four herb plants. You can start your herbs from seeds in the same pots where you plan to grow them or you can start them in seedling pots. You can also buy established herb plants at local gardening centres. One note: herbs are very sensitive when it comes to transplanting. It is probably best for beginning gardeners to start the seeds in the pots where the mature plants will remain. Some of the easiest herbs to grow are rosemary, oregano, chives, basil, cilantro and parsley. However, most herbs are easy and forgiving plants and should prove no problem to the dedicated indoor gardener. Plant your herbs as you would any other plant and water as needed. Many herbs, such as rosemary and oregano, are Mediterranean plants and prefer a drier soil. To test to see if your herbs need water, poke your finger into the soil around the plants as far as it will go.

If the soil at the bottom of the hole is dry, your herbs need water. Watch your herbs carefully for signs of drying. If the leaves begin to look wilted or become too dry, then you are not giving your herbs enough water. Herbs need very little fertilizer. Use a fertilizer formulated for house plants and follow the label directions carefully, applying the fertilizer only once every two or three months. Fresh herbs are a delightful addition to any kitchen. They give off a fresh scent and add great flavor to your cooking. And they are so simple to grow and maintain that anyone can have an indoor herb garden.

SOME COMMON HOUSE PLANTS CAN BE POISONOUS

When decorating with indoor plants, it is a good idea to know which plants can be harmful to pets and children. The following is a list of common house plants that should be kept out of reach of pets and children who might be prone to chewing or putting the leaves in their mouths.

Philodendron

The Philodendron is an ivy-type plant with heart-shaped leaves and is one of the most popular house plants. Eating or chewing this plant will cause a burning sensation in the lips, mouth and throat and may cause a rash. Rinse the child's mouth thoroughly with clear water. A large amount must be eaten to cause a severe reaction however it is a good idea to call poison control if your child does ingest the leaves.

Pothos

Pothos is another very popular house plant. It is a large ivy-type plant with heart-shaped leaves that are streaked with yellow or white. Eating or chewing this plant may cause a burning sensation in the lips, mouth and throat and may cause a rash. Rinse the child's mouth thoroughly with clear water. It does require a large amount to cause a severe reaction but it is a good idea to call poison control if your child does ingest the leaves.

Dieffenbachia or Dumb Cane

This is a large plant with large, oblong leaves that have white splotches. Eating or chewing this plant may cause a burning sensation in the lips, mouth and throat, a rash and it may also cause temporary speech impairment. Wash the child's mouth with clear water and call poison control.

Azalea or Rhododendron

This is a flowering shrub that comes in a variety of colors. This is a dangerous plant and should be kept away from children and pets. This plant may cause burning in the mouth, vomiting, diarrhea, headache, skin prickling, dimness of vision, muscular weakness, a slow heart beat, abnormally low blood pressure, convulsions and coma. This plant may cause death. If your child ingests any of this plant call poison control immediately.

Oleander

This is a flowering shrub with white, pink or red flower clusters, with long narrow leaves up to 10 inches long. This plant is very toxic and should be kept away from children and pets. This plant may cause death. If your child ingests any of this plant call poison control immediately.

PLANT CARE

Indoor plants add color, texture and warmth to the home. They allow year-round access to gardening and can even improve air quality. Many houseplants are easy to grow, but they must be given appropriate care in order to thrive.

Since your plants were probably started in a greenhouse — grown under ideal conditions — moving them into your home takes a bit of adjustment on their part. Proper watering and lighting are the most important components of indoor plant care, but humidity and temperatures also play a role. The trick is to try to mimic the climate of the place that plant came from.

Tropical plants thrive in warm, humid environments, while cacti and succulents prefer hot, dry climes. Of course, your home can't be everything to every plant, but you can take plant needs into consideration when choosing plants. And, with a few tricks, you can convince your green friends that they are living in their ideal environment.

PLANT SELECTION

The first thing to consider when selecting a houseplant is where you want to put it. Then match the space and lighting with the plant's requirements. Do you have a big spot by a sunny window or a small space with moderate light?

Next ask yourself if you are looking for a plant with beautiful green leaves or would prefer a flowering plant. Some flowering houseplants are seasonal while others will bloom year after year.A third consideration is how much time you can devote to a particular plant. A spider plant will take almost any amount of care (or neglect), while an orchid requires significant tender, loving care.

INDOOR PLANT CARE

Water

Potting soil should be kept moist, but not wet. Of course, there are always exceptions — succulents, and other thick-leafed plants do best when the soil dries out between watering. If the soil is kept too dry or too damp the plant's roots will begin to die, which can lead to inadequate growth or even death of the plant. Note: The Digital Moisture Meter (shown here) includes a handy plant care booklet and watering guide for over 150 plants. Use to prevent over and under watering by measuring moisture at the root level.

There are several methods to determine when a plant needs water. If the potting soil becomes lighter in color or cracked, it's probably time to water. Pick up your plant and gauge the weight after watering. After a few practice lifts, you'll be able to tell if the plant needs water just by picking it up. Of course, you can always stick a finger in the soil to determine how moist it is below the surface. For large plants, a hand-held moisture meter may be your best bet to determine how much water is present around the plant's root mass.

Dehydration

Do NOT let plants get to the point where they are wilting or the soil is pulling away from the edge of the container. These symptoms indicate dehydration and at this point the plant is already seriously stressed and the roots may be damaged.

Signs of underwatering include:

- Slow leaf growth
- Translucent leaves
- Premature dropping of flowers or leaves
- Brown, yellow or curled leaf edges

Overwatering

Too much water is just as detrimental as too little. Frequent watering forces air from the soil and opens the door for root-killing bacteria and fungus to move in. Overwatering is the number one killer of houseplants.

- Signs of overwatering include:
- Fungus or mold on the soil surface
- Mushy brown (maybe stinky) roots at the bottom of the pot
- Standing water in the bottom of the container
- Young and old leaves falling off at the same time
- Leaves with brown rotten patches

Watering on Demand

Plants requiring more water Plants requiring less water

– Flowering plants– Plants potted in clay pots– Plants grown in small pots– Actively growing plants– Plants located in direct sunlight– Large-leaved or thin-leaved plants– Plants that are native to wet areas. – Resting or dormant plants– Recently repotted plant– Plants grown in high humidity– A plant located in a cool room– Plants potted in non-porous containers– Plants with thick or rubbery leaves– Plants grown in a water retentive mix

For those who are too busy to keep up with a regular watering schedule, which requires checking individual plants every 3-4 days, there are several self-watering devices available. A moisture wick draws water from a dish of water into the root ball of your plant. Capillary mats and moisture tents also keep plants watered. You can always make your own self-watering plant container out of a 2-liter pop bottle.

Water Quality

Room temperature tap water should be fine for most indoor plants, even if there is chlorine or fluoride added to your city's water. Plants especially love rainwater or melted snow (unless you live in a region with acid rain). Avoid continuous use of softened water, which may contain sodium.

How to Water

Plants can be watered from the top down or bottom up. When watering from the top, try not to wet the foliage, while ensuring the entire soil mass is moistened. Water should be coming out of the drainage holes in the bottom of the pot. Tip: Be sure to dump any standing water from the saucer one hour after watering. If you prefer to let your plants do the work, set the plant in a dish of water and the roots (and capillary action in the soil) will pull up whatever they need. This method, known as bottom-watering, is a more thorough, if time-consuming, way to water plants.

Drainage

Good drainage is essential to healthy houseplants. Start with a good, organic potting soil (not regular soil) that has been mixed specifically for indoor gardening. Choose a container with drainage holes, or put a layer of pebbles in the bottom of a container without holes. The point is to not let the plant stand in water. From time to time, check that the drainage holes have not been clogged. And always empty standing water (don't run it back through the plant's soil).

Light

As with watering, every plant has different light requirements. Many plants prefer direct sunlight, but this may be hard to get inside a house. Placing a plant in a window might offer enough light, but some houseplants will need supplementing from a grow light.

Flowering Plants

Flowering plants generally do best in moderately bright light and for this reason windows located on the south, east or west side of the house are best for potted flowering plants. (African violets prefer north-facing windows.)

Foliage Plants

Foliage plants can be divided into three categories: those requiring low light, moderate light and high light. A dimly lit room should suffice for those few plants willing to survive in low light areas. Moderate light-needing plants will prefer a north-facing window, light diffused through a thin curtain or daylight without direct sun. Indoor plants that prefer high light will need to be in a south-facing window or under a grow light. Some plants will benefit from being moved outside in the summer to get a little extra light. Read about Moving Plants Indoors & Outdoors here.

Temperature

Many houseplants thrive in temperatures between 65-75° during the day and 55-60° at night. Of course, temperature preferences vary from plant to plant with tropical plants liking temperatures around 90° (or higher) and other plants growing better in cooler temperatures.

Humidity

Most plants thrive in high humidity — around 80%. Unfortunately, most homes are much drier, especially in the winter when forced heat can even further drop the humidity. Using a humidifier can help, but there are other ways to increase the moisture in the air near your plants. A small tray containing pebbles and water can boost local humidity as can grouping plants more closely together. Daily misting of the plant's leaves can help as well. For some plants, such as gardenias and orchids, keeping them in a bathroom or the kitchen (both usually have a higher humidity) can help.

Fertilizer

Every time a plant is watered nutrients leach out of the soil. Even if that didn't happen, plants would quickly deplete the nutrients in their soil. Unlike plants living outside, houseplants don't have a regular source of nutrient replenishment unless you fertilize them regularly. (Newly purchased plants have been heavily fertilized in the greenhouse and can wait a few weeks before getting started on a fertilizing regime.)

Fertilize once a month when plants are flowering or growing. During the winter, when plants are dormant or generally not growing much, fertilizer can be withheld. Note: The Indoor Plant Food (shown here) contains *NO* animal by-products or chemical anything — just pure, slow-release, organic goodness! If a plant is dropping its lower leaves, showing weak growth or an overall yellow-green color, it may need more fertilizer. It might also need more light or less water, so take the time to analyze all conditions before pouring on more plant food. Adding fertilizer when a plant does not need it can be worse than doing nothing at all. Tip: If a plant is wilted, water well first then apply a fertilizer later — after it has recovered.

Fertilizer Types

Choose an organic fertilizer specific to houseplants and read the instructions carefully. While natural fertilizers are less likely to burn or harm your plants than a synthetic fertilizer, it is important to apply the correct amount. In general, plants grown in low light will not require as much fertilizer as plants grown outside or in bright light.

To start, use about 1/4 the amount of fertilizer recommended on the label once a month. Then, if overall plant color becomes lighter, increase fertilizer applications to every 2 weeks. On the other hand, if the new growth is dark green, but the leaves are small and the space between the leaves seems longer than on the older growth, fertilize less often.

Tip: Soluble salts from synthetic fertilizers can build up over time and create a crusty layer of salt deposits on the soil surface. Remove this layer and leach the soil every 4-6 weeks with generous amounts of water to help avoid toxic salt build up. Excessive salts can damage roots and make the plant more susceptible to disease and insect attack.

Repotting

If your plants are thriving and growing the way you want them to, eventually they will need a bigger pot — or some fresh potting mix. Repot plants in the spring when they are just starting to grow. Vigorous root growth will allow the plant to adjust to its new container quickly. When it comes time to repot, choose an organic soilless medium made specifically for potting houseplants (maybe even specific to your species of houseplant). There are many to chose from, or you can make your own. Choose a pot that is bigger than the current container, but not huge. A pot that is too-big can encourage root rot and other problems because the soil will remain wet for days, or even weeks before it can be used by the plant. Take care with the root system when repotting to avoid damage. Carefully firm the soil around the root ball without compacting the soil. Leave enough space at the top of the new container for water and water thoroughly.

INDOOR PLANT CARE AND HOME ENVIRONMENT

Indoor plants help create a pleasant home environment. Small plants add color and scenery to windows or tables, while larger ones soften and blend with groups of furni-ture. As a part of the "indoor landscape," plants create a cool, spacious feeling, even in the warmest weather.

If you have been puzzled as to why some indoor plants thrive while others refuse to flourish, this publication is for you. Most indoor plant problems can be easily corrected and even more easily prevented. First, realize that most of the foliage plants known as house plants are native to tropical areas. This is why, for instance, most indoor plants prefer a humid atmosphere and indirect light. Of course, there are exceptions, and these are also discussed. Keep in mind that sound cultural methods, preventive care, knowing your plant's requirements, and careful attention are the best substi-tutes for a green thumb.

LIGHT

This publication includes a partial guide to the light requirements of some common indoor plants. Use it as a guide for the selection and placement of your plants. Plants vary considerably in their light requirements. For example, plants such as the croton need direct sunlight, while philodendrons will grow under lower light intensi-ties. If plants are not receiving enough light, their leaves turn yellow and die. Most people depend on natural window light for the growth of their plants. Natural light may be adequate if plants are close to windows. However, the amount of natural light a plant receives, decreases dramatically the farther it is placed from the window or its source of light. Usually, plants must be located close to windows to receive enough light for them to grow and flourish. If window sill space is in short supply, hang basket planters, make glass shelves, or build a "bay window greenhouse" by extending the window and adding shelves for plants that require higher light intensities. Light from reading lamps is also beneficial, but the new "broad spectrum" fluorescent lights are usually better. Generally, artificial, fluorescent light 10 to 14 inches above the plants will provide sufficient light for plants requiring medium light intensity, such as the pothos. A combination of one warm white to one cool white light is best. If your plants appear spindly, they need more light, should be closer to artificial light, or your fluorescent light bulbs are too old to produce the proper wavelength for growth and should be replaced. For even growth, turn the plants once a week so they receive light on all sides.

Temperature and Ventilation

Most indoor plants grow well between 60 and 75°F. They may become spindly if kept warmer. Always keep plants away from hot or cold drafts, warm appliances, and heat registers. Flowering plants will retain blossoms longer if lower temperatures are provided. Temperatures above 75°F hasten the death of flowers and make plants spindly as well as less resistant to disease and insect attack. Some plants will thrive in hot and dry conditions. Proper ventilation is necessary for good plant growth. In most homes, ample fresh air is provided. However, guard against escaping fumes from gas appliances or furnaces.

Humidity

Most plants require a higher humidity than that of the average home. Any means of increasing humidity will be beneficial to your plants. Today many heating systems are provided with humidifiers, which should be kept full of water. Sprinkling or syringing plants with water is not effective as the water evaporates rapidly. Fittonia (Fittonia verschaffelti), baby's tears (Helxine soleiroli), and strawberry begonia (Saxifraga sarmentosa) will tolerate high humidity conditions. Growing plants to-gether in a terrarium or setting them on a tray of gravel with moisture in the tray are good ways to raise humidity.

Watering

Improper watering is the cause of most house plant problems. Both under- and over-watering can cause leaves to yellow and fall. Check plant soil daily to see if they need water (soil dry 1/4 inch down and tapped pot sounds hollow). If needed, add water until moisture drips out of the drainage hole of the pot. Wait a few minutes and water the plant again until moisture drips out of the drainage hole. Be sure to discard drainage water. Do not allow the bottom portion of the pot to stand in water. With small plants, weight is a good indicator of when water is needed. Dryer plants feel lighter than those with moist soil. If you prefer, water from the bottom. Place the plant in a saucer of water until the top of the soil is moist. Then remove the plant and let excess moisture drain away. Never keep ordinary indoor plants standing in water continuously. A well-drained soil mixture is essential for ease of watering. Heavy soil mixes hold too much water and often cause the plant to rot at the crown. When watering, the water should be visible on the surface for not more than a few seconds. If it tends to stand on the surface, mix perlite or sand into the soil to improve drainage. Note: Water all plants from the top once a month to prevent mineral or salts build-up, and be sure to water any plant before it wilts. A build-up of white or brown crusty material on the soil or rim of the pot indicates salt accumulation.

Fertilizing

Water your house plants with a dilute fertilizer solution, especially during the summer. Prepare a solution by mixing one teaspoon of soluble fertilizer in 1 gallon of water. Such fertilizer analyses as 20-20-20, 5-10-5, 4-12-4, or 7-7-7 may be used. Apply once a month during the growing season. Often a grower will purchase a commercially prepared soluble fertilizer. He should then use it according to the directions on the container. Slow release fertilizers save time. Follow the label directions. Apply amounts (usually given in teaspoons or tablespoons) to the soil. Each time you water, some nutrients are released. A new application can be given when recommended on the label. If you have fertilized plants through the summer months, little or no fertilizer is necessary during the winter. Plants should be allowed to become dormant at this time and produce little growth.

Soil Preparation

Adequate water drainage is essential for plants growing in containers. Ensure good drainage by mixing soil with liberal amounts of sand or perlite. These materials help aerify the soil, improve root growth, and allow water to easily penetrate the mixture.

A good soil mix includes two parts peat moss, one part sand, and one part perlite; or one part silt or clay loam garden soil, one part organic matter (peat moss), and one part coarse sand or perlite. Commercial potting soils can also be used. These usually require the addition of one part sand or perlite to improve drainage and plant growth. Sterilize the soil mix to remove weed, insect, and disease organisms. Place the soil mix in an oven and heat to 180°F for 20-30 minutes. Tools and containers should also be sterilized. Wash equipment in a chlorine bleach-water solution (1:10 by volume) to kill disease organisms.

Repotting

In the spring when new growth starts, turn each of your plants upside down, tape the edge of the pot, and remove the plant. If the roots are in a solid mass, they need to be repotted. Shift the plant to a pot (with a drainage hole) 1 or 2 inches larger in diameter. Fast-growing plants may need an even larger pot. Add new soil to the container. For most plants, use 1/3 sphagnum peat moss, 1/3 garden soil (loam), and 1/3 sand or perlite. For cactus, use 1/2 quartz sand and 1/2 garden soil. After adding soil, water as usual. You may need to repot fast-growing plants more often than once a year. Usually you will not need to water as frequently for a few months after repotting.

When an indoor plant has reached a desirable size, do not repot it. Instead, remove some of the soil at the top of the pot at least once a year and replace with fresh soil.

Additional Care

Most plants, except those with hairy leaves, respond to an occasional bath. Syringing the plants occasionally with a fine spray of clean water removes accumulated dust and keeps the stomates (pores) open.Some broad-leaved plants, such as palms, may be cleaned by rubbing with a soft cloth. Never use an oiled cloth. However, special leaf waxes, available in many florist shops, may be used on glossy-leaved plants without causing harm.

Use a fine, soft brush, such as camel's hair, to clean hairy-leaved plants. Never use leaf wax on these plants. Climbing plants may be trained to make them attractive and easier to care for.

Summer Care

Many house plants can be invigorated by growing them outside during the summer. Remove all dead leaves and prune where necessary. Place the plants in a location suitable for their light requirement, e.g., low light requiring plants should be placed on a shady porch.

For minimum summer care, dig a hole, put gravel or cinders in the bottom, and set the pot so the top is at the original soil line. Mulch with peat moss, and water. Usually these plants need watering more often than when indoors. During the summer, lift or twist the pot at least every month to prevent roots from growing through the drain-age hole in the pots.

Lift the plants with their pots before they are endangered by autumn frosts and repot if necessary. Inspect the plants for any insect or disease infestations before returning indoors. A shower with plain water or insecti-cidal soap may take care of most insect problems. Contact your local county Extension agent for further advice, if needed. Be sure to follow all directions when using any pesticide. Before you vacation, water all indoor plants well. Either have a "baby-sitter" water your indoor plants or move them out-of-doors as suggested. However, drying out can be delayed for a while if pots are sunk in moist vermiculite or peat moss and kept in a cool spot away from direct sun.

Bibliography

A B Menan: *Introduction to Plant Anatomy*, Rajat Publication, Delhi, 2008.

A.K. Jain: *Encyclopaedia of Control of Plant Diseases*, Anmol Publication, Delhi, 2010.

Amar Tyagi: *Plant Diseases (2 Vols-Set)*, Anmol Publication, Delhi, 2006.

Archana Singh: *Plant Diseases and Their Biological Control*, Aavishkar Publication, Delhi, 2011.

Aslam Khan: *Plant Anatomy and Physiology*, Kalpaz Publication, New Delhi, 2001.

Aslam Khan: *Plant Diseases*, Kalpaz Publication, Delhi, 2001.

B A Mannan: *Encyclopaedia of Plant Anatomy (4 Vols-Set)*, Anmol Publication, Delhi, 2008.

B. P. Pandey: *Plant Anatomy*, S. Chand Publisher, Delhi, 2001.

Balram Sharma: *Plant Tissue Culture and Transformation Techniques*, Daya Publication, Delhi, 2008.

Barrows, M.: *A Survey of the Intestinal Parasites of the Primates in Budongo Forest*, Uganda. Glasgow University, 1996.

Berger,Robert W: *In the Garden of the Sun King: Studies on the Park of Versailles Under Louis XIV. Washington,* D.C.: Dumbarton Oaks Research Library and Collection,1985.

Bradley,Richard: *New Improvements of Planting and Gardening.3d ed.London*: Mears,1719 and 1720.

Campbell,Susan.Charleston Kedding: *A History of Kitchen Gardening.London:* Ebury,1996.

Campbell,Susan.Cottesbrooke: *An English Kitchen Garden.London*: Century,1987.

Campbell,Susan: *Walled Kitchen Gardens.* Princes Risborough: Shire,1998.

Chakrabarty, Falguni: *Adaptation of the Santals to the Hill-Forest Environment*, APH, Delhi, 2012.

Chambers,Douglas: *The Planters of the English Landscape Garden: Botany, Trees,and the Georgics.New Haven, Conn.,and London:* Yale University Press,1993.

Chase,Isabella Wakelin Urban.*Horace Walpole: Gardenist.An Edition of Walpole's "The History of the Modern Taste in Gardening," with an Estimate of Walpole's Contribution to Landscape Architecture*.Princeton,N.J.: Princeton University Press,1943.

D.A. Evans, W.R. Sharp, P.V. Ammirato and Y. Yamada: *Handbook of Plant Cell Culture: Vol. 1 to 3*, Panima Publication, Delhi, 2009.

Endress: *Plant Cell Biotechnology*, Springer Publication, London, 2004.

Evelyn,John: *The Compleat Gard'ner.Translated from the French Instructions pour les jardins frutiers et potagers by Jean-Baptiste de la Quintinye,1690*.London: Gillyflower,1693.

Gilbert Wooding Robinson: *Soils : Their Origin Constitution and Classification*, Biotech Books, Delhi, 2005.

Gupta, O.P. : *Water in Relation to Soils and Plants : With Special Reference to Agriculture*, Agrobios, Delhi, 2002.

Howitt,William: *Rural Life of England.3rd ed.London: Longmans,Brown,Green,and Longmans*,1844.

Loudon,John Claudius: *Encyclopaedia of Gardening.5th ed.London: Longman, Rees, Orme, Brown, and* Green,1835.

Mountain,Dydymus (alias Thomas Hill): *The Gardener's Labyrinth.New York.London: Garland*,1982.Facsimile of 1577.

Prasad, T.V.S. : *Soil Chemistry : Nutrient and Water Management in Agricultural Soils*, Dominant Pub, Delhi, 2009.

Reynolds, V.: *Budongo: A Forest and its Chimpanzees*, London, Methuen, 1965.

Svieking,Alber Forbes,ed: Sir *William Temple upon the Gardens of Epicurus with Other Seventeenth-Century Essays*.Gollancz,1902.

Thomson,Robert: *The Gardener's Assistant,Practical and Scientific.1st ed.6 vols*.Glasgow,1859.

Index